ON BECOMING A BIOLOGIST

Also by John Janovy, Jr.

Keith County Journal
Yellowlegs
Back in Keith County

ON
BECOMING
A
BIOLOGIST

John Janovy, Jr.

PERENNIAL LIBRARY

Harper & Row, Publishers
New York
Cambridge, Philadelphia, San Francisco
Washington, London, Mexico City
São Paulo, Singapore, Sydney

A hardcover edition of this book is published by Harper & Row, Publishers, Inc.

First PERENNIAL LIBRARY edition published 1986.

Library of Congress Cataloging-in-Publication Data

Janovy, John.
  On becoming a biologist.

  (Harper & Row professions series)
  "Perennial Library."
  Bibliography: p.
  1. Biology—Vocational guidance. I. Title.
QH314.J36   1986   574′.023   85-42573
ISBN 0-06-091363-0 (pbk.)

86 87 88 89 90 MPC 10 9 8 7 6 5 4 3 2 1

CONTENTS

ACKNOWLEDGMENTS

There are three men who are mentioned often in the discussions that follow—Dr. Harley P. Brown, Dr. J. Teague Self, and Dr. Leslie A. Stauber. Drs. Brown and Self are retired from the University of Oklahoma faculty; Dr. Stauber is deceased. These are the people who did the most to shape my career as a biologist, and in appreciation, I would like to devote a few comments to each.

Harley Brown was my major professor for my master's work at OU. He was a person who lived for his thoughts. He had an immense amount of knowledge at his mental fingertips but felt no need to demonstrate it except when asked. I've known few teachers who placed less significance on academic politics than Dr. Brown. His world was that of the invertebrates; Machiavelli did not exist. My first contact with Harley Brown was a moment of good fortune. I had returned to campus after a time in the army, determined to be an ornithologist. I had a degree in math and eleven credits of zoology, including the freshman course. G. M. Sutton, my hero at the time, would not accept me as a graduate student. Dr. Richard Goff, the summer adviser, must have had some kind of insight; he recommended I take a course called "Natural History of the Invertebrates." The textbook was Pennak's *Freshwater Invertebrates of the United States;* the instructor was Brown. Before the summer was over, I waded in what seemed like every creek in southeastern Oklahoma.

In the fall, I enrolled in protozoology. The textbook was Kudo's *Protozoology;* the instructor was Brown. Before the semester was out, Dr. Brown had offered me A Problem on the biology of the carnivorous ciliate, *Dileptus anser.* It was not until I had graduate students of my own that I understood what an act of gentle faith that offer was. At the time, the chance to watch *Dileptus* become monstrous on a strange diet was the one thing I needed most to validate my decision to become a biologist.

Two years later, M.S. in hand, I approached J. T. Self, who was, at the time, my wife Karen's employer. Self was also a parasitologist, a former chairman of the department, and, to those who didn't know him, a frightening bear of a man. Although Dr. Self did not seem to be, himself, an expert in all the areas in which he was supervising graduate work, I was impressed with his students; they seemed an unruly bunch, taken with fish guts and worms, shotguns, seines, and old microscopes. Their problems had a cosmopolitan flavor; their work, and the way they talked about it, smacked of The Grand Challenge.

Self accepted me as his graduate student, and then, when he realized that I had no idea what I'd gotten into, recommended I take his course in parasitology. If this seems somewhat backwards, it is only a reflection of the approach to education that characterized the Zoology Department at OU at the time. The faculty didn't seem to have a great deal of respect for convention. They studied what they wanted to study and worried little if at all about the practical applications. Furthermore, they were not about to let themselves be bound by a bunch of rules they knew would be confining. Self was probably the most independent of the bunch.

I learned a number of lessons from J. T. Self, all of them of great significance in my attempt to become a biologist. First, you didn't whine because there was no money. Instead, you drove your own car, used your own shotgun to collect, and studied what problems nature would allow. Second, a microscope that was good enough for Ronald Ross to solve the malaria life cycle in the 1800s, and thereby receive a Nobel Prize, was damn well good enough for a grad student in 1960. After all, it wasn't the microscope that made the scientist, it was what you chose to look for with it. Third, you didn't get all undone by disappointment or failure. Instead, you shrugged it off and went on about your work.

Dr. Self not only expected his students to enjoy the hell out of parasitology, he also insisted they be socially graceful, mature in matters of academic politics, proud of their profession, and above all honest with themselves. These were high expectations. We tried to meet them. I can't speak for the others; I know I enjoyed the hell out of parasitology. In the twenty years since I graduated from OU, I have come to realize how truly rare is the kind of intellectual climate that J. T. Self and the University of Oklahoma provided.

Leslie Stauber was my post-doctoral supervisor at Rutgers. From him I learned, or at least began to sense, just exactly how much labor was required before a person could truly say he'd discovered something about nature. Stauber was a world figure, deeply respected by all who knew him. But he treated me as if I were a colleague on his own level. I suspect he knew that his example would serve as a humble reminder that complex biological problems are great equalizers.

There is a long list of others who have contributed, unknowingly, to this book. These have included students, foreign scholars, ecologists, molecular biologists, biochemists, immunologists, and philosophers. Most have been close acquaintances; some have been failures, others successes; all have taught me things that have nothing to do with living organisms and everything to do with ways to pursue intellectual endeavors. Most I respect. Their lessons, taught by example, have always taken their place within the professional context established for me by those three men characterized above. It is for this that I would like to especially thank my three great biology teachers.

—JOHN JANOVY, JR.
February 1985

ON BECOMING A BIOLOGIST

1.

NATURALISTS

> Don't be an ornithologist if you
> can help it. But if you can't help
> it, go ahead. . . .
>
> —FRANK CHAPMAN

1.

It was that idyllic time between Little League and cars with girls. We lived on the far edge of town. In an afternoon you could set traps, marvel at the studied placement of possum tracks, and grind the iron red Oklahoma clay into your very soul. Parents were people who drank whiskey and played cards. Their friends tried to test my manly inclinations by asking:

"What are you going to be when you grow up, Johnny?"

"A naturalist," I'd reply. They'd smile. Even then I could sense their thoughts: There's no such thing as a naturalist anymore.

Only my grandfather was sympathetic. He'd seen passenger pigeons, known the old west, and watched Indian Territory become the forty-sixth state. His clothing and demeanor revealed a certain longing. His house was filled with books of natural history. I remember sitting for hours staring at a color plate of men spearing a woolly mammoth. The smell of cigar smoke still takes me back to those days, those books.

"We were born a hundred years too late," he'd say. I never thought then of the nearly sixty years that separated us. We seemed of the same generation, at least in our minds.

Since those times humans have placed a biological experiment on Mars, sent the songs of whales on an odyssey outside the solar system, passed a city ordinance against gene-splicing experiments, bought stocks in gene-splicing companies, added words such as "Love Canal" and "Three Mile Island" to our American lexicon, and put personal computers with prey-predator models in the elementary classroom. The way things are going, I'd have to say my grandfather was wrong. We were born a hundred years too soon.

Developments in science since the Second World War have been stunning. There are people alive today who have seen science fiction become reality. The accelerating pace of science and its offspring, technology, has forced us into consideration of questions that in earlier times were the domain of philosophers: Who are we, where are we going, is there an identifiable human nature, is our research immoral? Had I been born a hundred years earlier, my birth would have coincided with another naturalist's epic journey: Charles Darwin would just have finished his voyage on the *Beagle*. In 1837, my parents would never have accepted the idea that a young naturalist's trip could alter our vision of life on Earth. In our time, however, discoveries that assault our senses are familiar events. Yet there remains a shared awareness that confirms the implications of Darwin's world view—namely, that the human species cannot live apart from the planet upon which it evolved. We share a common bond with even the most bizarre beetle of the Peruvian rain forest. A belief in that common bond might, in fact, be the most fundamental characteristic of a biologist.

Scientific, unlike religious, belief is derived from evidence and is subject to modification. Yet belief functions in biology, and in science in general, in much the same way it functions in religion—to direct behavior and maintain values. To truly believe in a common bond with the Peruvian beetle is to hold the insect in high esteem. We respect the bug because of what it represents: life itself. The values that are upheld by a biologist's beliefs are decidedly noncommercial ones. For example, the worth of an organism is found in its contribution to our understanding of life, not necessarily in our ability to convert it into money or prestige.

Humanity as a whole does not seem to share the values of a biologist. This difference is not surprising when you consider the following: Most biologists perceive the human species as only one of the more recent of millions that have occupied the biosphere over the last three and a half billion years, and therefore have enormous interest in the nonhuman components of the natural world.

This view of humanity as a late intruder—perhaps the price of becoming a biologist—is in stark contrast to the values associated with other professions. Attorneys, physicians, businessmen—all are consumed by human activities, conflicts, desires. The collective introspection that marks our species is enforced by these professions. A biologist studies "nature," however, and in doing so inevitably comes to regard humanity as the most effective competitor for the world's resources. The conclusion must be tinged with admiration: No other species' accomplishments approach humanity's accelerating cultural evolution, which in essence represents an escape from the "restrictions" of organic change. But the conclusion must also be tinged with sorrow, for no other species seems to possess the power to destroy overnight what cannot ever, anywhere in the universe, to our knowledge, be replaced. To become a biologist today is to adopt and live with this set of conflicting realizations.

Values are, of course, cultural phenomena. A great body of literature deals with the mechanisms by which such traits are acquired. Briefly summarized, that literature suggests: (1) There are times in early childhood when we are especially receptive to certain environmental influences; (2) early experiences often shape our later behavior; (3) it is not always possible to determine which early experiences translate into which adult behaviors; and (4) once established, certain adult behaviors are exceedingly difficult to change.

Biography seeks to unravel the mystery of life choices and show us by example the effects of experience on our tracks through time. The biographies of biologists always tell of events in which nonhuman organisms are held in high regard. On her deathbed my own mother reflected on her life and talked of times that must have been among her finest. Startlingly young, with an equally young husband off to his first job, she was too broke to go anywhere in a strange land and had a child to entertain. So she did the only thing she could do for free. Every day she put me in a stroller and walked to the Audubon Park Zoo in New Orleans. Who knows what she said to me, if anything? Maybe she just let me look until I was satisfied, then moved on. It would be pure romance to attribute adult career choices to such specific early experiences. But today I cannot pass a zoo without stopping. I stand in front of cages and look until I'm satisfied, then move on. As a senior in college, I encountered a teacher who also took me for a walk, let me look at animals until I was satisfied, then moved on. In May I was to receive a bachelor's degree in mathematics. In March I decided to become a professional biologist.

That teacher was Dr. George M. Sutton, a world-renowned ornithologist who was, at the time, Research Professor of Zoology at the University of Oklahoma. Although I will later address the relationship between teaching and research, as a biologist's life is by definition tangled up in both, for now let it stand that this holder of the highest title a university can bestow, Research Professor, was also a teacher who could direct career choices by example. The message was obvious to all who encountered Sutton. Here was an exceptionally articulate, broadly educated person, an artist who drew upon the poetry of Gerard Manley Hopkins as a teaching device for advanced classes, who was dead-eye with a shotgun, who lived a rich and challenging life, and who was adored by the public. He destroyed any stereotyped image of "biologist." He possessed all the traits, physical and mental, we would like to give our heroes. But overriding them all was his single unshakable belief that, of all the world's resources, none had higher value than birds.

Sutton's own earliest memories included a remarkable incident. His father had taken him walking on a bridge over the Mississippi at Aitkin, Minnesota. A blackbird or grackle landed on the railing. Sutton's father lifted him up, left him standing alone on the railing above the brown and swollen river, and pointed out that the grackle, perched in a similar position, had no fear of falling. George Sutton later claimed the experience was one of two clear and genuine childhood memories in that it had never been mentioned later by his parents. His other genuine memory was of a man who had a collection of bird skins, certainly a unique and impressive phenomenon for Aitkin. One is almost reminded, by these tales, of the metaphorical power of a primitive society's rites of passage. Courage and individuality became symbolized for Sutton, when he was at a very malleable age, by birds. A towering figure at the age of eighty, Sutton continued to express his own courage and individuality through his study of birds.

It's impossible in a book as brief as this to make a complete case for early experience as the crucial factor in life choices. It may also be impossible to account for the effects of a lifelong career as a biologist on the strength of childhood "memories." Especially in the case of prominent, thus in our eyes successful, people, their view of what the public wants to read may influence their choice of words for the introductory paragraphs of autobiographies. Yet, again and again, biologists claim their earliest memories are most often not of humans. Darwin's recollections of the furnishings in his dying mother's room are followed quickly by those of collecting natural items and attempting to identify plants. Charles Darwin described himself at the age of nine as "a naturalist." To support his adult work he had a habit of almost indiscriminate collecting. Writing his biography at the age of seventy, he obviously felt it important to convey this sense of a life spent gathering items from nature.

Biology today, of course, demonstrates how truly monumental is the legacy given to us by naturalist Darwin. Molecular biologists, biochemists, cell physiologists, behaviorists, and ecologists are all able, and sometimes willing, to place their work in an evolutionary context. Geneticists by the very nature of their research are evolutionary biologists without trying to be. While the work of all these scientists may have profound implications for humanity, it is doubtful that much of it is undertaken specifically for that reason. Most of it is done instead to satisfy personal curiosity, or to answer questions that just seem, at the time, to need answers.

Of special interest in this discussion of early life experiences are those of biologists who cannot, as practicing scientists, rightfully be called naturalists. Might even a gene splicer find his or her origins in a childhood fascination with a butterfly? Does a biologist who is captivated by the intricacies of protein synthesis have a certain eye for park squirrels and their acorns? Evidence suggests they do, or at least claim they do. At the age of ten, Thomas Hunt Morgan, father of modern genetics and 1933 Nobel Prize winner, may have been, with his collection of stuffed birds, eggs, insects, and fossils, indistinguishable from the pre-adolescent Darwin. Jacques Monod's father, an avid reader of Darwin, instilled in young Jacques an early interest in biology. And Marshall Nirenberg, 1968 Nobel Laureate who deciphered the genetic code, wrote his M.S. thesis at the University of Florida on the ecology and taxonomy of caddis flies.

Obviously not every potential biologist possesses so clear a set of values as George Sutton, Charles Darwin, and Thomas Hunt Morgan. But the models suggest that values are a legitimate tool, probably in the same category as electron microscopy: that is, they allow you to work in areas, especially areas of thought, you would not have access to otherwise. Like other tools, they may be acquired, although one may possess them and not realize it until confronted with the fact. In retrospect, I had probably always been a biologist, but when I enrolled in George Sutton's class I was trying very hard to be something else. When people later asked why I came back to graduate school, I told them, "Sutton showed me it was all right to be a biologist, after all." What he actually showed me was that the values I held were legitimate ones. The values then asserted themselves and directed my actions.

In the years since that semester with Sutton I've encountered thousands of students, many of them majoring in biology. I could count on one hand the number who were willing or able to discuss values in the same way they could discuss the Krebs cycle. Yet they all must have had some values. I wonder now if we did all those students any favors by structuring a curriculum that emphasized the latest knowledge in biology. They bought textbooks and studied someone else's summary of information that was by then at least five years old; we discussed journal articles to present newer techniques and discoveries. Then, to teach them about why people do things, we sent them to the humanities departments. Did we expect them to integrate the knowledge to arrive at a sense of why people did biology? Yes. But did we try, or do we try today, to assist that integration by straying from the subject of someone else's latest experiments to talk about why that journal article was important to us, or might be important to them, personally? No.

So in one critical area—the *reason* biologists study living organisms our whole lives through—education is left largely to chance, and the responsibility for those lessons falls on student shoulders. The idea that science classes must, from bell to bell, deal only with observations, interpretations, and experimental design is a delusion. Original science cannot ever reveal, except in the most indirect way, why the teacher who presents it so articulately made his or her life choices. Those who ask, "Should I become a biologist?" or, "Am I a biologist without knowing it?" thus have the task of ferreting out their role models' values.

2.

But what does the foregoing discussion of values mean for the person who today considers a decision to *be* a biologist? First, it suggests that such decisions may be quite different from decisions to *do* something. We begin to sense in our role models a force that in the end tells them what they are, rather than what to do. The metamorphosis from "I do . . ." to "I am . . ." is a critical one. It not only frees a mind to explore the world in a way that is not available to an intellectual larva, but it is also ultimately required for the production of original work in any area, not just biology.

Unfortunately, biologists, like attorneys, must buy groceries and find shelter, and metamorphosis into a state of "I am" is strongly inhibited by concerns about employment. Inspired by uncertain economic and political climates, the entire movement urging young people to go to college in order to find a satisfying and financially rewarding job is a movement that undermines the decision-making power of a would-be biologist. To retain that power you must recognize your values for what they are—namely, the elements of identity. A job, on the other hand, is a task you do for someone else. You can do a job without being something. But being requires an active commitment to sustain identity.

The activities that biologists use to sustain their identity often have no apparent relationship to the familiar world. But biologists do mature, find financial support somewhere, melt into the general population, and surface in the most unexpected places. Such as a local high school gym. In a fit of charity, I agree to sell tickets at a junior varsity girls' volleyball tournament. The woman who works with me fills the time between customers with casual acquaintance talk: Where do you work? . . . etc. Finally her eyes take on a look that I've come to see as a warning.

"I know one of the people in your department. He lives over in our neighborhood. His wife seems real nice, but he . . . ," she says.

"Oh?"

"He studies frogs in South America." There follows a considerable evaluation of people who study frogs in South America. Throughout this description of my colleague it is clear that a fascination with Neotropical amphibia is a pursuit she cannot reconcile with some image of a productive member of society. As we sit there taking money, I get my periodic look at life outside the ivory tower. It is amazing what adults will do to try to get into a junior varsity girls' volleyball tournament free. But what is more amazing is the task my colleague might face in trying to convince the world there is nothing wrong with studying Neotropical frogs. Finally the woman asks about my own research. I tell her of some studies on parasites that are capable of infecting humans. This work is clearly legitimate. I pass the test as a fellow member of the species.

But on the way home, I review my own sense of what my colleague is. First, he is an exceptionally intelligent, well-educated, and multilingual person who works exceedingly long hours at a noble profession, that of the teacher. He publishes numerous research papers every year, the result of his original study and thought. His name is known in certain circles around the world. He travels to exotic places at least once a year: the Peruvian Andes, Amazon jungles. The animals he studies are a collection of living jewels. He can knowledgeably discuss history and philosophy and will try to discuss anything else. He loves to fish. He and his wife have contributed two bright children to our society. He lives a rich and rewarding life. He wrestles daily with the problem of conveying a sense of impending ecological disaster to the public. He wouldn't trade places with anybody. And to this gentleman, there are very few things more important than Neotropical frogs.

But of what value to society is his work? Will additional knowledge of the Leptodactylidae provide a cure for cancer, eliminate pollution, end the threat of nuclear war? No. What it will do is contribute to our overall picture of the planet upon which we live. In recent times, it has been shown that the tropics are the repository for the bulk of the genetic information that spells life on Earth. Current research brings into harsh focus the speed with which tropical forests are being cleared and the frightening long-term implications of such clearing. Fifty years ago we didn't equate the clearing of tropical forests with elimination of life itself. Molecular biology had not given us the vision of strange species' genes as common property in which we all had an interest. But today, the work of naturalists such as my friend is beginning to show us just how much of this genetic information exists. Through tropical studies, the world has been shown to be far richer, and more complex and fragile, than we imagined it to be. My friend's research helps form the fundamental concepts that should shape our decisions about the kind of relationship we must have with the planet. And he surrounds himself with frogs.

The decision to become a biologist demands an attachment to the world of living organisms. There may be several million species of plants, animals, fungi, protists, and microbes. The vast majority of them are insufficiently studied. They are all distinct in some way, and thus reflect several million different relationships with Earth. They all have evolutionary histories and relatives. Most have been here longer than humans. Their relationships with one another add a level of complexity above that of their own specific structure and history. One of the major accomplishments of biology has been to demonstrate how little we actually know about all these organisms, and by implication, our own life support system. Organisms are, or are constructed from, cells. The complexity of cells is at least as great as that of ecosystems. Thus, whether we look beyond ourselves or within ourselves, we still see a world which we are only beginning to understand. There is much to be done by biologists. The exotic lives of people like my herpetologist friend are waiting to be repeated over and over again by succeeding generations. But they are not likely to be repeated by people whose values are focused only on humans.

3.

What do biologists actually see? How do these scientists perceive the world? The questions need to be answered if only because all professions are characterized by their typical world visions. Individuals whose perceptions are inconsistent with these accepted visions are outcasts. Such people usually become either miserable and ineffective burnt-out cases, or else they foment rebellions that redefine the typical world view.

The perceptions that are shared among professionals are used as the basis for defining accepted areas of inquiry within a profession, and to determine legitimate problems and their solutions. This general phenomenon is the subject of Thomas Kuhn's classic work, *The Structure of Scientific Revolutions.* Kuhn begins his discussion by presenting the concept of the paradigm. Paradigms are intellectual environments, frameworks within which we work. The nature of a particular environment is determined by leaders, or those with reputation, by the state of the knowledge, and by available technological tools. These all operate to establish the limits of investigation and define valid and important problems. For example, the problem of how to clone a gene was not a recognized one prior to the rediscovery of Mendel's laws of inheritance. In 1880 you could have asked the question, "By what mechanism is DNA synthesis regulated?" and no one would have cared. Professional biologists of the time had neither the understanding nor the equipment to consider such a question. Instead, their knowledge and technology would have established an environment—i.e., a paradigm—in which other questions were important. An apprentice's paradigm is generally consistent with the world view of the professionals. The beginner is taught what the professionals know, feel is important and think can be studied. Thus, the paradigm both directs and limits intellectual development.

Scientific revolutions occur when paradigms are overthrown and new questions become valid, important, and potentially answerable. Scientists are in the business of trying to make discoveries that will change our working environment—that is, our paradigms and world view. It is for this reason that Kuhn is virtually required reading for anyone who would be something defined by intellectual activity—for example, a biologist.

Of course no one has asked a broad and extensive cross section of biologists how they see the world. An individual's sense of the common vision is a result of years of contact with other people who have claimed to be biologists, who have made life decisions to become biologists, or who are biologists without knowing it. One day I find myself sitting at a corner table over beer and pizza with two people. One of them is an exceptionally bright first-year graduate student, a person I always mention when predicting success and fame. The other is a young woman who will soon be entering graduate school. Neither is worried about a job. Both have done some research, taught laboratories, attended scientific meetings. I ask what they see when they walk across campus.

The young man says "birds," but then goes on to explain

how he's always either watching the ground or the sky, which I translate into "robins" or "gulls." He concedes that between the ground and the sky, he averages straight ahead. The young woman sees processes. She always knew there was a beginning and an end, but now she sees "what goes on in between." Her big course this semester is cell biology. Both recognize some essential differences between themselves and their fellow students. Both see their environment in terms of their identity. Both already consider themselves biologists.

Differences between world views become glaringly apparent when a biologist, as teacher, walks into a university classroom of nonmajors. For months, often years, before this happens, departmental committees have asked, "What should we try to teach these people?" "These people" are typically a collection of up to several hundred eighteen- to twenty-year-olds who have already, perhaps on the advice of a high school counselor, made what they believe are life decisions. Few if any of them hold values that include a place for the leptodactylids. When asked at the beginning of the semester to describe what the world is like, the vast majority will respond that it is filled with war, politics, education, and money. If questioned about their choice of majors, most will give an answer that translates into money. Few see the living world so obvious to my two companions over pizza and beer.

So with what equipment should the biologist approach the mob? Obviously, among the list of intellectual weapons should be the profession's world view. If it is not expressed, then the biologist will not bring to society his or her essential contribution: one version of the truth. We will then produce another group with biology on the transcript but not much that passes for biology on the mind. Such a group will have little potential for making rational decisions on major issues involving the life of the species. If a student leaves biology class with information, but without the vision to see beyond the surface and into structure, process, complexity, and dependency, then we have not actually taught biology. Ideally, the teacher should convert students into at least partial biologists.

Contemplating the responsibilities of an instructor is a useful exercise for the serious student considering his or her role in society as a biologist. The task of teaching a subject brings into focus a person's own thoughts about what is significant enough to be passed on to succeeding generations. Deciding what is to be taught by a biologist as opposed to, say, a chemist or linguist helps us choose what we feel is of fundamental importance to us in our specific role. Sooner or later, directly or indirectly, every professional will address the matters of role, responsibility, identity. Biologists are not immune to this experience.

There is a mental exercise that I sometimes use to help people either acquire a biologist's world view or to recognize it in themselves. If done honestly, it forces them not only to look at their surroundings differently but also to combine seemingly unrelated ideas. It generally works like a charm. The task is to discuss the biological content of the magazine *Art in America*. (Although virtually any reasonably sophisticated art magazine will do, this particular publication is both heavily illustrated and of high production quality.) What one discovers is the use of plant and animal subjects by nonbiologists, and although the use is symbolic, it often extracts some essence of the organisms portrayed that would escape an unromantic person. For example, a recent, and typical, issue illustrated, among other biological subjects, a botanical garden as backdrop to a Miró sculpture, Giacometti's *Dining Table with Gilded Leaves, Frogs, and Sparrows,* Alexander Hogue's *Desert Mesa in the Big Bend,* which could have substituted for any photograph of the desert biome in any freshman textbook, Jack Thompson's *La Rosa del Cardinal* (voluptuous torso with head of a cardinal), an André Masson abstract with identifiable fish, Rudy Fernández' surreal *Courting Disaster,* with a perfectly executed trout, a veritable bestiary in Harry Koursaros' *Pandora,* and Gallery Hirondelle's advertisement of the sculptor Mihail's "spiritual safari to free the Elephant from the shackles of Man's dominating ambitions" and the casting of his first of eleven life-sized bronze elephants. Once they've been asked to look, and told where to look for it, most students can generate pages of discussion on this kind of material.

The exercise is beneficial in that it generates an unexpected amount of biological awareness. First, for a biologist to ask a student to read an art magazine is an affront to the student's sense of order. Second, the assignment clearly has nothing to do with biology as the subject is perceived by the student. At this point the student's perception of biology is simply narrow. But just by being forced to look beyond a familiar context, the student acquires a measure of sensitivity to biological content in unexpected places, and thus makes some progress toward acquiring a world view similar to that of my companions above. Obviously, if you exercise this kind of sensitivity easily and naturally, on your own, you have one of the common properties of people who would be nothing other than biologists.

4.

To this point we have considered the values, visions or world views that biologists, as opposed to other professionals, hold. These values and visions will not necessarily guarantee a successful career in biology, but they will validate a career decision long after a student has left formal academic training, at a late time in a person's life when careers often need validation. It is a common experience among all professionals, not just biologists, to wake up one day and ask, "Is being a . . . really what I want to do with my remaining years?" At such times values and visions will prove their worth, for they provide the best foundation for serious consideration of one's contributions and potentials. At the end of his career will my herpetologist friend renounce the Leptodactylidae as worthless and obscure? Did Sutton on his deathbed decry a life wasted on birds when he could have been an equally articulate attorney? No.

Values and visions also have another significant role—namely, that of recognition of oneself. A certain value set allows you to see yourself as a biologist, and not as an attorney, banker, or used car salesman. Such self-recognition can be the factor that turns a searching student into a professional with a mission, allows an older person to add richness to a wandering life, and helps anyone to place disappointments and frustrations into their proper perspective. I think that self-recognition is what I developed in Sutton's ornithology class. The fact that I am not now an ornithologist is testimony to the good judgment of some other teachers as well as to the diversity of intellectual opportunities for biologists. But I do consider myself a biologist. And I have never looked back on that decision made without a great deal of thought one day in March.

I did have an opportunity, however, in the six years after that decision, to be closely associated with a number of people whose naturalist values were, like Sutton's, virtually non-negotiable. In every case they pursued their intellectual work as if it were of monumental importance, and in so doing validated every day my decision to become a biologist. Harley Brown (see Acknowledgments) seemed at the time to be able to identify every organism, from the nearly microscopic to distant soaring hawks, that we encountered on field trips to southeastern Oklahoma. Furthermore, he became excited over most of them, even though I had the feeling that he'd seen them all dozens of times on previous expeditions. He was the first grown man I'd observed who simply waded into a stream in his field clothes and began turning over rocks, exclaiming his discoveries in a voice that compelled a student to take note. By the end of my first field day with Dr. Brown, a sense of the earthly richness a naturalist's eye could discern had started to take shape in my mind. To Sutton's birds, Brown added the rest of the animal kingdom, although his emphasis was on the free-living forms.

J. T. Self, with his group of students, showed me the parasites. Two graduate students in particular played a key role in my decision to study parasites—Dan Harlow and Jerry Esch. Both had been in the ornithology class in which I'd decided to become a biologist instead of a mathematician. By the time I returned from the army, they were well into their graduate studies. Dan was asking a grand question: Do pectoral sandpipers transport parasitic worms from one continent to another on their migrations? Jerry's research was, at the time, a model for physiological studies that were done within the context of a species' life in nature. His question was a rather sophisticated one: Do larval and adult tapeworms utilize glucose according to the same metabolic pathways? Dan's work took him to the field to collect sandpipers; Jerry's involved jackrabbits and dogs. Both spent long hours in the lab processing their materials, but they never forgot that the worms lived in, and had come from, nature. Dr. Self seemed to enjoy sitting back and watching both of these young men struggle with their research.

Dan used to come over to our little rented cottage two or three nights a week to sit around and talk about science and scientists. Jerry was enrolled with me in Brown's advanced invertebrate zoology course. In the middle of lab, we'd take a break, go up to Jerry's office, drink terrible instant coffee, and review the structure of worms. With this kind of exposure to the naturalist's values, and recognition of myself as a person who was, down deep, a biologist, there was no question that I would accept Self's offer of a research assistantship on a large project involving the intercontinental movements of viruses and parasites.

The project was based on the assumption that organisms

that migrated from one continent to another could carry parasites, including viruses of medical and veterinary importance, and in the process serve as a source of potential pathogens for resident animals. In practice, however, the work required that I commute from Norman, Oklahoma, to the marshes at Great Bend, Kansas, where the bulk of the Central Plains field work was to be done. My home away from home was a small building put together by a multitalented gentleman, H. A. "Steve" Stephens, the project's resident collector and the last of the old-time naturalists. Whatever doubt I may have had about the worth of the naturalist tradition in biology, Steve laid to rest. His personal residence was a camper on the back of a pickup truck. Inside there was a pet prairie dog, ample reading material, a desk where he did his writing, and a small kitchen. Steve was working on a book about the woody plants of Kansas. His reasons for living in a camper seemed obvious: When a particular shrub is about to bloom, and one needs to draw a picture of the flower, it helps to be able to simply drive up to the plant and wait for the blossoms to appear.

Over the course of the next three years, I was to become completely immersed in an attempt not only to do project work but also to answer my own thesis question: How does the life of a host affect its interactions with a parasite? My studies concentrated on two species of birds and a single species of parasite, *Plasmodium hexamerium*, which each host contracted from mosquitoes. The two birds, meadowlarks and starlings, lived under very different circumstances—the starlings were tree-hole nesters, the meadowlarks stayed in the open fields. I had embarked upon a journey of discovery that led eventually to the tops of rotted cottonwood trees, the inner stillness of a prairie marsh, the flooded Kansas pastures with mosquitoes so thick one could hardly breathe, and the Gulf Coast of Texas. In the end, I did learn something about the transmission of bird malarias, but not nearly as much as I had expected in the beginning. More important, I discovered I was in the company of naturalists all parading as professional biologists.

Dr. Self probably had the most perspective on a scholar's life of any of my teachers. He was a very practical man, a no-nonsense problem solver and decision maker. His stint as department chairman had eliminated any fear he may have had of campus politicians, his contacts with people outside the department helped him to appreciate the philosophers, his exposure to faculty from other universities inspired him to send his doctoral students on to post-doctoral training in the programs of renowned scientists, and his sense of wonder at nature kept him working and publishing well into his seventies. But of all his experiences I think one stood out as a beacon. His first graduate student was a boy named Bob Kuntz, a kid who loved snakes and was a collector in the Darwinian tradition. Self spoke often, and somewhat smugly, of the difficulties Kuntz had in getting admitted to graduate school. I could see why my teacher had such special feelings about his first student. By the time I encountered J. T. Self, he was near retirement and Bob Kuntz was one of the most widely published and well-known figures in his field.

Among those who were fellow biology graduate students during my years at OU can now be counted college presidents, deans, department chairpersons, taxonomists, ecologists, behaviorists, biochemists, businessmen, real (not campus) politicians, teachers, wildlife artists, and photographers. None of them trained specifically for these various roles in society, although they may have emphasized particular areas of study by taking selected courses. Instead, I prefer to think they all evolved, intellectually, as a result of their fascination with organisms. A dozen years after leaving OU, I found myself in need of the lessons I'd learned from these people outside the classroom. Campus wheels had turned at my institution; machinery and reductionism seemed to be replacing the curiosity about nature that I knew should characterize the profession; and in our rush to become modern we'd thrown away our respect for plants, animals, and microbes. What values must one hold in the technological age? I asked. The answer came as a book I wrote entitled *Yellowlegs*, a quasi-biographical piece of vision-quest literature that celebrated the naturalist tradition among those people who guided my attempts to become a biologist.

5.

Is a decision to become a biologist one faced only by a college student? Is it necessary to become a professional in order to be a biologist? In both cases the answer is no. There is a wealth of opportunity for serious and rewarding study by "amateurs." While nonprofessionals are usually limited by lack of technical training and sophisticated equipment, and by the incredible expense of some types of biological research, they have access to almost unimagined resources in the world of nature. There are so many kinds of organisms that can be studied with a magnifying glass, so many known only by their structure and not their habits, so many whose lives are more easily analyzed by the ingenious mind than by a gas chromatograph, that biology should not be restricted to professionals. But ultimately a middle-aged widow's decision to become a biologist will rest on the same questions as a college freshman's: Do I have, or can I accept, the values and visions that open up this rich and complex world?

Of course, the concept of an appropriate vision is also relevant for the professional engaged in more formal—i.e., scheduled—research, where one sees not only problems to be solved but also questions to be posed. A skillful scientist is one who first asks the right questions, who can "see" such questions in the surrounding biosphere. Right questions tend to be those that can be answered using known or devised techniques, that serve as case studies, analogs, or metaphors for larger questions, and that generate additional questions with similar characteristics.

A fundamental property of a mind that asks the right questions is that it not only sees complexity of structure and function within organisms but also sees how this complexity is superimposed on uniformity. The uniformity is that of cell structure and function, the structure of DNA, the processes of gene expression and evolution, the phenomenon of interdependency. For the biologist, complexity reveals itself in the specific manifestations of life processes. This vision is easily demonstrated with simple examples: There are a quarter million ways to express the concept "beetle," ten thousand ways to express "grass," and nearly twenty thousand expressions of "orchid." A true biologist walks through town and sees complexity upon uniformity in the sparrows digging through last night's garbage, reads the paper while the words "tiger . . . lizard . . . lily" leap out from the comic page as if they were the real thing, and sits in a crowded restaurant and watches *Australopithecus*, *Pongo*, and *Pan*.

The world of a true biologist is also filled with potential investigations. This person goes home in the evening and despairs for only seconds over the bindweed strangling his tomato plants, for in that bindweed he recognizes a lifetime of experiments. He may pick an underground stem and take it into his laboratory for no other reason than to think for a few days about how this material can be converted into understanding of the processes that govern plant growth and development. Perhaps a "right question" will come into his mind. For example: What biochemical events occur when a white stem is stimulated to produce a green leaf upon exposure to sunlight?

And finally, in this matter of vision, a biologist sees interdependency. The possible relationships that can exist between organisms are often summarized by introductory textbooks as (1) prey-predator, (2) competitive, and (3) symbiotic. These categories imply that organisms either eat one another, compete for various resources, or live in some (often) obligate relationship in which one species depends directly on another at the individual level. But prey-predator relationships also involve dependency, that of the predator upon the prey. Since competition can drive evolution, competitors may depend on one another, in an almost metaphysical sense, for their identity. The relationships that are established within a community of organisms therefore provide organization and structure, regulate populations, and give a community a "life" of its own.

Relationships also dictate the specific paths by which energy and materials flow within a community and ecosystem. It is a fundamental principle of life on Earth that energy has a net flow in only one direction, while chemical elements are recycled. Thus life as we know it depends on a continuous supply of energy from the sun and the availability of materials which may have already been used in living organisms. Although the statistical chances are minuscule, this principle allows that a carbon molecule in your body today may have been incorporated into the DNA of some lumbering dinosaur. The energy that powered the beast, however, is not available now, nor will it ever be again.

The extent to which biologists are aware of these fundamental principles of the science, is the extent to which they are aware of interrelationships, potential and otherwise, in their daily lives. My two pizza-eating friends see not only robins, gulls, and metabolic pathways along the campus walks, but the potential interactions among them. Robins eat worms, which are converted into CO_2 by Krebs cycle reactions; the CO_2 is incorporated into leaves of campus plants which give off oxygen that is breathed by the gull to help power it to the Gulf of Mexico. Nothing in that sentence, from its implications to its almost cell-like complexity, is strange to a biologist.

Finally, the combination of dependency and complexity gives rise, in the biologist's mind, to diversity. In a walk through town or woods, a biologist observes the kinds of organisms, be they cultivated or wild, and notes their relative numbers and the niches they seem to occupy. From block to block the diversity may change: A vacant lot offers a mini-course in ecological succession; a garden easily reveals the unstable nature of monoculture; trees tell the age of a neighborhood as easily as they tell the year a southern farm was abandoned.

Today's biologist will also extend his or her analysis to the dominant species: Are there niches for humans, do societies obey the same principles as the organisms with which they surround themselves? New York is structurally complex, a virtual, albeit synthetic, equivalent of a tropical rain forest, populated with layered communities, guilds, and a diversity of intellectual species occupying their own niches. Such niches are not necessarily present in the "deserts" of our society, but a biologist strolling down Madison Avenue knows there are thriving, adapted communities in the Kalaharies, the alpine tundras, and the taigas of our culture as well. These are not the metaphors that occupy the minds of attorneys, bankers, engineers. Or, if they are, we have some biologists chafing within common disguises.

6.

The field of biology that a student encounters today is stunningly different from the one Darwin experienced. A hundred and fifty years ago science wrestled with not only the mysteries that have confronted humans at all times, but also with a philosophical climate in which religion was a powerful and ever-present force. Although scientists proclaim intellectual neutrality, social climates can never be eliminated as an influence on the practice of a profession. Thus, while Darwin's contemporaries did science in an era when religion influenced the interpretation of data, yours do science in an overpopulated, weapons- and health-obsessed society which acts as if the world has an unlimited supply of fossil fuel. The times affect science as much as they influence the art, music, and literature a culture consumes. During the second half of this century, two inescapable factors cast their shadows on every scientist's work: (1) monumental problems to be solved if our species is to remain extant, and (2) technological tools of such power that the public equates their use with science. The former, the problems of *Homo sapiens* (mainly overpopulation), I will consider later. The technological tools need to be dealt with now, for they have a bearing on the image I have been drawing of "naturalist."

What, specifically, are these tools? First, they are machines, pieces of equipment. Your familiarity with their names is as sure an indicator of the times as your calendar: Scanning Electron Microscope, Transmission Electron Microscope, amino acid analyzer, gas chromatograph, scintillation counter . . . personal computer, main frame . . . laminar flow hood, ultra centrifuge. Second, they are standard analytical approaches that allow machinery to be incorporated into biological investigations; examples are techniques such as cell fractionation, nucleic acid and protein purification, *in vitro* methods for everything from enzyme assays to cell-free protein synthesis, cell culture, and hormone assays. Third, they are the mathematical equations and statistical methods that allow scientists to convert raw observations—that is, numbers—into proofs of hypotheses. As with any set of tools, it matters less that you possess them than that you use them well.

Because the public today tends to view science in terms of its tools, there is a subtle danger that biology students (who, after all, emerge from the public) will find it easy to confuse technology with science. Given the proper equipment and training, it is not difficult to make monoclonal antibodies. What is infinitely harder is to walk through the woods and decide, on the basis of your observations, against which antigen you should make monoclonal antibodies. Until you are able to make this decision, or perhaps more important, until you are willing to try, you will remain a technician instead of a biologist. To make the transition, however, you must acquire the mind-set of a naturalist. If you are successful in doing so, your apprenticeship is likely to end, for you will begin to apply your tools to the raw material of organisms' lives to develop understanding.

It is important for every practicing biologist to ask: What effect do these tools have on the way I approach organisms, biology, and science in general? At the most obvious level they influence thinking itself, if for no other reason than by their constant reminder of what can be analyzed. The ever-growing range of available tools should also serve as a reminder of not only what training it will take to become a certain kind of biologist, but also what may happen to you after having made a commitment to a line of study built around a certain tool. This may be a critical consideration for students. Advanced math, chemistry, and physics, which are almost essential today for minimal technological expertise, are not easily acquired by an established professional, so that a decision to become a biologist almost obliges one to a certain minimum exposure to math and physical sciences at an early age. Yet there are only so many hours in a day, and using them in the acquisition of tools inevitably takes time away from the study of organisms. In the end you may find that formal education has provided you with the biologist's tools but not the naturalist's values and world view. This would not be so bad if the history of science did not show that tools grow dull quickly. High-tech biology compounds the dilemma by its adoration of vogue.

We should, then, soberly address the question: What has high technology done for and to biology? The answer is: Plenty. It has expanded, and is continuing almost daily to expand, our understanding. We see in organisms things we never saw a generation ago. Today a parasitic worm such as *Schistosoma mansoni* is viewed in terms of its specific ultrastructure, metabolic pathways, fatty acid composition, and the antigens that it uses to cover its own unique proteins and thus "hide" from its human host. Or, a single species of nematode, *Caenorhabditis elegans*, is viewed as something that will finally help us discover the principles by which embryonic development is regulated.

Furthermore, our knowledge now enables us to recognize interrelationships among observations. We look at a plant and see a spectrum of secondary compounds. A generation ago we didn't understand the role these molecules played in plant/animal interactions. Today we see plants defending themselves against herbivory, using secondary compounds as repellents, and we wonder if the gene splicers can put the ability to manufacture certain secondary compounds into food plants, thus eliminating our need for commercial insecticides. At this point the flippancy of the phrase "high-tech vogue" disintegrates.

The above example is worthy of closer examination. What exactly would it take to provide humankind with a strain of corn that could repel, through its own secondary products, the insects that now compete effectively with us for food? What would the payoff be? The second question is fairly easily answered. An enormous financial burden would be lifted from agriculture and a major source of environmental toxins—i.e., pesticides—could be eliminated. And we can speculate on the answer to the first. It would require identification of those compounds that actually function as insect repellents in plants (organic chemistry), demonstration that some of these compounds are effective specifically against corn pests (experimental design, statistical analysis), identification and description of the metabolic pathways by which the compounds are produced in nature (biochemistry), identification and isolation of the genes responsible for the pathways' enzymes (molecular biology), successful insertion of the required genes into the corn genome (genetic engineering), and production of the seed grain itself (agronomy).

Given the nature of biological science today, this potential accomplishment does not seem outlandish; the basic elements of this scenario are fairly familiar to us, even from the popular press. There is very little in the above paragraph that has not already been demonstrated to be possible. But professional biologists, especially those with modern research experience, would be less sanguine than the public regarding the actual likelihood that biology could soon eliminate our need for insecticides. Basically, what remains is to accomplish the feat of combining parts of organisms, or, more precisely, combining specific parts of specific organisms. Other successful combinations are already generating commercially useful products.

Ecology and ethology are the modern and mature versions of natural history. Ecologists and behaviorists have thus tended to acquire tools that help them understand the activities of whole organisms, rather than single genes, proteins, or metabolic pathways. These tools are primarily mathematical ones. Math, probably more than chemistry or physics, influences the thinking of those who employ it. Mathematics provides one with highly structured processes which can be used in an almost infinite number of ways. Having used math, however, and appreciated its powers, one is reluctant to retreat into a weaker mode of communication, and instead steps forward to open another intellectual door. In the next chamber lies abstraction, the representation of process by marks on paper, and a stairway to places where the purest thinkers live. Not many biologists climb the stairs. But all are familiar with math's utilitarian cousin, statistics.

Today, because calculations and statistics are done largely by computer, it is imperative that a student whose curiosity leads to ecology and behavior become computer-literate. "Student," in this case, does not apply only to the college junior. One of the most powerful experiences I have had recently is that of watching my colleagues wrestle with computer literacy. We are talking here about tenured senior faculty who suddenly see the future staring them in the face.

The University of Nebraska operates a summer field program at a place called Cedar Point Biological Station. One summer a new faculty member borrowed from the city campus, for a period of five weeks, one Apple II personal computer, and set it up in the main laboratory. The college students were reluctant users. Unable to write their own programs, they felt embarrassed. Confrontation with a new language and its characteristic grammar and syntax was not anticipated to be a part of a course entitled "Vertebrate Zoology."

The students' embarrassment, however, was nothing compared to that of the teachers. A twelve-year-old faculty child with a short elementary school course in computers sat down at the machine and produced wonders: graphics, original games, records of his kitchen employment. By the end of the summer, the faculty members knew they had no choice but to come to terms with what they had witnessed. Some bought personal computers from out of their own pockets. Others began immediately writing grants. The next summer every course used faculty-written software, and there were four personal computers in camp. I've just described a mini-drama in which a group of professionals and students were shown what it takes to be a naturalist in today's intellectual climate.

Computer literacy, hardware, and software are subjects upon which any person can now get more advice than he or she could have ever, in a wildest dream, imagined. In a field in which equipment, programs, and languages are multiplying at the rate of bacteria, it is less important to choose the right system than simply to choose something. Every computer language requires that you communicate in a highly structured and logical way. It is this fundamental type of communication with which you must become familiar; the specific machinery and vocabulary are secondary.

In fact, it is common for people who are computer-literate to be comfortable with more than one language. Not long after that twelve-year-old child showed professionals their future I had the privilege of serving on a search committee to select candidates for a faculty position in ecology. All of the top twenty or thirty candidates listed their ability to use several computer languages as part of their qualifications. Furthermore, in doctoral programs all over the country, computer skills are becoming accepted as equivalent research tools in place of the foreign language requirements of earlier generations. Linguists tell us that language affects the way we think. If this is true, then our next generation of computer-literate naturalists is acquiring not only a means of manipulating numbers, but also a highly structured and logical way of thinking about biology.

Chemistry, physics, and math are clearly, in their own right, discrete areas of intellectual endeavor. But they are also ones that the complex science of biology has borrowed from in order to answer questions about organisms. Historically, human progress has been characterized by a blurring of the lines between disciplines, not only in so-called "hard" sciences but also in the social sciences, arts, and humanities. In most cases, however, the overlaps involve technical contributions rather than values and visions. Artists may choose space-age materials, but these materials are then used to present an artist's, not a scientist's, world view. The same can be said of the relationships between a biologist and chemistry, physics, and math. A molecular biologist may have to convince reviewers that a research paper's chemical and statistical techniques have been done correctly, but the published paper is usually much more interesting to biologists than to chemists.

Interactions between fields can be thought of as occurring through a filter of values and visions. That these interactions occur is an element of your daily working environment as well as a fact of history. Not only will you have to deal with them in order to become a biologist, they will influence the kind of biologist you become. The most successful biologists, however, tend to keep the filter intact. For example, it is easy to become vitally interested in problems of enzyme regulation because they must be solved before you can explain the manner in which an organism deals with environmental change. On the other hand, in order to study enzyme regulation you may have to become as skilled in chemical techniques as a professional chemist. Research problems can consume decades of intense work. If, after those years of research, you are more interested in the enzyme than in your original question about physiological adaptation, then your values and world view will be those of a chemist instead of a biologist.

7.

When complementary fields exchange values and visions instead of simply techniques, intellectual communities tend to respond with controversy. Perhaps the best recent example of this phenomenon can be found in the reaction to sociobiology. Sociobiologists interpret human behavior within the context of the broad field of ethology, seeing in humans general behavioral patterns that are common to the "lower" animals. Familiarity with this field is important to a potential biologist because E. O. Wilson, the founder of sociobiology, made biology a direct human concern in a way even the layman can appreciate.

For example, in sociobiology literature humans are described as "mildly polygynous," a primate social characteristic that is predicted by differences in body weights between males and females. "Mildly" implies that the human male, on the average, consorts with three to five females, instead of twenty or thirty, as might the males of some other mammalian species. Culture, however, directs the specific manifestation, but not the fact, of this trait. It is not unusual for a male to have two wives and be helped by three secretaries over the course of a career in business in this country. A sociobiologist would interpret this situation as a rather natural one—that is, having a biological basis. If you are a female reader, you can easily see from this one example how the findings of sociobiology can generate controversy. Wilson does not state that social structures that have a biological basis should necessarily be maintained. What he does imply, in essence, is that a biologist has as much to say about the human condition as the politicians.

Let us imagine that "biologist" is the cultural equivalent of a family or genus, that your specific interest is the equivalent of a species, and that your individual approach to that interest is a variant within the species definition. Society therefore becomes the environment which will select for or against your cultural phenotype—i.e., the form your individual education and skills will take. If, like most students, you are interested in employment, you will try to discover those cultural phenotypes for which society is selecting. If you are interested only in perpetrating your original ideas regardless of your state of physical and financial well-being, then you will probably try to discover cultural niches where your mutant ideas are not selected against. If the sociobiologists are correct, in either of the above cases you as a job seeker will be forced to take into account the extent to which *Homo sapiens* is a social species, for social behavior dictates the structure of much of the working environment.

As a result, you will conduct your career in a world in which group attitudes determine the fate of individual ideas. No matter what your contributions, you will eventually have to find employment. Whether you get a paying job in government, business, or academia, you are still likely to encounter dominance hierarchies, lingering sexism, cultural conservatism, religion or equivalent belief systems, and intergroup rivalries. I hope that in the next few paragraphs we can deal with this situation in a serious, noncynical manner. Humanity will not change its basic nature before you receive a Ph.D. Or, for that matter, before you retire. It is best to look at ourselves objectively and plan accordingly.

A close look at group behavior reveals that social forces foster opportunities as well as inhibitions. If you understand the mechanisms by which these forces work, then career hurdles which seem difficult can suddenly become easy. If you are planning to teach, then recognize that a classroom is fundamentally a social unit; your teaching will become significantly more satisfying. For example, you may have to establish a dominance hierarchy with yourself at the top in order to counter dominance behavior, by a very few class members, which keeps a bright but shy student from making a significant intellectual statement. You may also have to devise ways to thwart group judgment in order to preserve a student's sense of self-worth so that he or she can progress. Thus, when you understand social behavior, teaching becomes a challenging and rewarding contest between your own problem-solving powers and a social order that might otherwise appear fixed.

Similarly, in research, you will find that social forces tend to determine which problems are "good ones" and which solutions "will work." We are back to Thomas Kuhn (*The Structure of Scientific Revolutions*). As a beginning biologist, if you attempt to step outside your paradigm, then expect resistance every inch of the way. Negative criticism about technique can be answered with ingenuity. But you should also be prepared for the apparently damning "Why are you asking *that* kind of question?" I'm not dissuading you from trying to step outside your paradigm. I am saying that you should not expect it to be done easily. If you accomplish it, you'll be history's hero.

Social behavior will also influence your career through the funding of research. For example, it should be obvious that an agency with the name National Institutes of Health will have certain missions and vested interests. Virtually every source of money society will provide to support your intellectual activities will have some stated or practiced policy which can be thought of as a group identity. To receive funding, you will either bend your thoughts to the source's mission, or convince the source its vested interests are served by your work, or do without. Alternatives tend to fall in a writer's world: marry for money.

In the long run, the social sciences can thus be considered in the same light as chemistry, physics, and math—viz., as disciplines that influence the way you do biology. And in a very real sense, the social sciences can and should also be thought of as tools. Like the physical sciences and math, they enlarge our vision of what can be analyzed, opening up areas of study, stimulating the investigation of new questions, and redefining the word "biologist." Unfortunately, most universities have not yet come to this realization. Rarely, for example, will a semester of psychology substitute for a semester of physics in a biological sciences curriculum, even though the former might well be more beneficial than the latter, over the course of a career. So it may be largely up to you to educate yourself in the social sciences, to choose courses in history, political theory, and group behavior, and the faculty members who teach them, with the same seriousness and care with which you choose your physics, chemistry, and math.

8.

Among the items I inherited from my father, a petroleum geologist, was his introductory zoology notebook. He had taken the course in the early 1930s from a Dr. Richards, who, it turned out, was the major professor of Dr. J. Teague Self, the teacher who directed my own doctoral work. I always knew my father had graphic skills. His maps were works of perfection. He could do things with drawing instruments that were beyond belief, I thought even as a college student. So it came as no surprise when, not long after his death, I opened his zoology notebook to discover some pencil drawings. By this time I was a professional biologist. It was obvious from certain anatomical details, irregularities of placement of certain invertebrate organs, that the drawings had been made directly from specimens. They were also stunning works of art.

My father had been following a tradition that goes back hundreds of years: A biologist must be observant. The world of nature is complex, and generations of biologists have tried to make sense of this complexity by recording their experiences in drawings. There may be something in the profession that brings out the artist. Art has played an exceedingly strong role in the naturalist tradition. Exquisite botanical drawings dating back hundreds of years had a very practical purpose at the time they were executed—namely, as an aid in identification of medicinal, edible, and poisonous plants. Leeuwenhoek's introduction of the microscope as a valid research tool in biology was accomplished to a major extent by his drawings of things which had simply not been seen before the late 1600s. The same statement, substituting appropriate respective dates and the word "photographs" for "drawings," could be made about transmission and scanning electron microscopists. George Sutton's watercolors served, whether he intended them to or not, as a window through which the general public saw into the workings of a scientist's mind. His Arctic field sketches and Mexican paintings, exhibited in the university's art museum, drew shoulder-to-shoulder crowds. Louis Agassiz Fuertes, Sutton's mentor, produced paintings of such biological strength and artistic beauty that they remain unsurpassed today. "Audubon" is a household word.

The strongest evidence for the significance of art to a biologist today can be seen in professional journals. A comparison of any dozen biological journals with their counterparts in chemistry, physics, or math will reveal the extent to which biology relies on graphic communication. All scientific publications will have graphs as well as flow diagrams, charts, tables. But in the biological journals you will find detailed anatomical drawings, graphic depictions of behavior and ecological relationships, and photographs of biological materials. Electron microscopists are especially proud of their work; most are intensely competitive about the quality of photographs that appear under their names. They may be dealing with the intricacies of subcellular anatomy, but they communicate in the same manner as Sutton and Fuertes.

The world of life is, above all, the world of objects. Although biology is certainly not devoid of theory, our fundamental interest in organisms always has the greatest influence on our thinking. While we endure abstractions, we prefer to see things that are alive. Failing that, we like to see things that our minds can easily translate into living organisms. More often than not, this desire is manifested as dependence on a picture. The graphic richness of an introductory biology text, in comparison to a text in math or the physical sciences, is further testimony to our reliance on art, in all its forms, as a communication device. So it is perfectly appropriate for a lab instructor to ask a student to draw what is seen in the dissecting pan or through the lenses of a microscope. In a complex but tangible world, understanding often comes from combining observation of what is new with the physical act of graphic representation.

So in a very real sense, art is a tool. But the full breadth of art—including, for example, abstract expressionism, photorealism, and historical trends in architecture—is often less accessible to a potential biologist than are physics, chemistry, and math. Somehow the physical sciences seem legitimate adjuncts to a biologist's education, while art tends to fall into that nondescript category of "humanities electives." Furthermore, public schools tend to encourage an early separation of the arts and the sciences. Students who excel in one while struggling in the other will take appropriate refuge. The result is often an early commitment to courses of study, even in the public schools, in which those "interested in science" avoid painting, drawing, and ceramics like some intellectual plague. All this comes home to roost about the time you have to design and produce the plates for your first paper.

Thus the following suggestion is not too far out of line: Study art, fairly seriously. Begin with any good book on cave art, read the modern interpretations, speculations, on what these first naturalists were trying to convey. Note the context in which this art occurs, the evidence that the artists possessed detailed anatomical knowledge, and the values suggested by what is and is not represented. Visit art museums whenever you get the chance. Study the different ways single themes are presented down through the ages, in different media, in different social contexts. Give abstract art an honest try; ask what might have been in the artist's mind when the piece was conceived, how the work might have been modified by environmental forces during its ontogeny. Then budget some time to wonder what a freshman biology student does with "abstractions" such as transcription and translation.

What you will be doing with this activity is addressing the question: In what ways do humans express their perceptions of reality? "Reality" in this case is a complex and often misinterpreted world which includes everything—from images out of the depths of the artist's mind to the most seemingly mundane street scenes. The value of this exercise to a biologist is fairly obvious. At some time in your career you will ask: How do I best express the reality of my observations and interpretations, which may range from the depths of cellular processes to the seemingly mundane regulation of group behavior? With any luck, at this point the investment of time spent in museums will pay off.

9.

In the course of this chapter, I have tried to comment on the naturalist tradition in biology, the values and world view that biologists hold, and the tools that allow us to explore a complex and interacting world of organisms. I don't intend to imply that one cannot be a biologist without having attained the values and acquired the tools mentioned. There are professionals, in fact, who might take offense at the suggestion that they are part of a naturalist tradition. Some cell and molecular biologists fall in this category. Similarly, field men often treat the pursuits of molecular biologists with disdain. There is little to be gained from such provincialism, except perhaps some short-term benefits that accrue to those who dabble in academic politics.

There is, however, much to be gained from a broad and tolerant view of a chosen field. The most valuable advantage is one analogous to that held by the genetically variable species: In the long term, a catholic approach allows exploitation of numerous opportunities. Biology today is an exceptionally wide field of endeavor, almost as varied as the subject matter that supports it. Try as you might, you will not be able to keep up with all the latest advances and techniques outside your immediate areas of interest. A wide view, however, of what, in your own mind, is legitimate biology will help. Furthermore, it will enable you to see applications for your own discoveries that may not be obvious to a more narrowly focused person. The converse is also true; breadth will allow you to see how contributions in other areas can apply to problems that you may not currently be able to solve.

At one time or another, in order to see yourself as a biologist you will have to make some kind of original observations of nature and meld them into a synthesis of understanding. The increasing diversity of biological subject matter, coupled with growing interconnections between physics, chemistry, math, and the social sciences, provide many opportunities that were never dreamed of a generation ago. It is unlikely that this trend will be reversed. It will, of course, be difficult for you to maintain a generalist's approach to biology, if for no other reason than that modern science puts a premium on the specialist. But an extensive literature on careers and life tracks, ranging from the serious to the pop psych/soc, suggests that if you choose narrow and early specialization, then in later years you will become vulnerable to role extinction. Scientists today are caught in an intellectual paradox: To operate at the professional level they must confine their practice, while the field itself demands that they enlarge their visions. One of your greatest challenges will be to resolve personally this paradox. One of the surest ways to meet the challenge is never to lose the naturalist's sense of wonder at the world of organisms.

2.

THE PRACTICE

OF BIOLOGY

> If you study man by the method suited
> to chemistry, or even if you study him
> in the light of what you have learned
> about rats and dogs, it is certainly to
> be expected that what you will discover
> is what chemistry and animal behavior
> have to teach. But it is also not surprising
> or even significant if by such methods
> you fail to discover anything else.
>
> —J. W. KRUTCH

1.

What is this work we call biology? How is it actually done? Does it bear any resemblance to what we remember from high school or freshman lab? The last question is easiest to answer: No. Most of the biology has been done for beginning students by the time they enter a lab to do a formal exercise. On the other hand, serious amateurs may already have done more biology than they realize. The amount of biology done by advanced students is in direct proportion to their experience at truly independent investigation. Teachers eliminate most of the biology from classes in the interests of economy. Academicians grow frustrated over the unwillingness of sources to fund their best biology while giving them thousands of dollars for projects that will "produce results." You can probably tell from the foregoing that my picture of biology as it's actually done is likely to be rather uncompromising.

The practice of biology involves a multitude of activities and decisions. The activities range from the manipulation of biological material to the development of analytical techniques, the search for financial support, the design of experiments, the construction of hypotheses, the preparation of manuscripts. The decisions range from relatively easy ones (which techniques are best for a particular problem, where to set up glassware) to more difficult ones (which of several preliminary hypotheses are actually testable, how is this grant proposal to be phrased) to the most difficult decision of all—what to study.

For the individual, the practice of biology depends primarily and directly on this most difficult of decisions. If you are unable to decide what to study, you will never study anything seriously. This statement seems deathly trivial. Unfortunately, because it reflects a fundamental characteristic of our public educational system, it is anything but trivial. In general, the public school systems in this country work very hard, albeit unconsciously, to deprive students of their power to make intellectual decisions. For example, few students, especially younger ones, decide for themselves which historical events to read about, which systems of government to analyze, or what literature is appropriate for people of their age. All of the forces of organization, standardization of curriculum, and, especially at the elementary level, insistence on proper behavior, inhibit intellectual decision-making development. The teacher tells you what to study; you learn it. Something in the air suggests that if you learn it well enough to get good grades and do well on ACT and SAT tests, you will earn a scholarship and/or admission to a fine university. The complex human brain extends this image of grades = life success to include financial rewards, marital happiness, and career satisfaction. As any person over forty knows, none of this is necessarily true.

The unfortunates wrestle long days which sometimes fade to years with the choice about what to study. No one can tell another person how to decide the answer, but one can offer some useful suggestions. Above all, it is crucial to recognize that science is a human activity, as human as art, music, literature, athletic competition, and war. Human intellectual activities are driven by either desire or unidentified internal forces. Science demands a sustained interest and effort that last over decades. Without desire, sustained effort is impossible. To be a biologist you must study organisms because you want to. Although you may be able to come up with dozens of other reasons for use in cocktail party conversation, such explanations are likely to be rationalizations put forward to satisfy the needs of nonscientists.

With the recognition of desire as a driving force, we begin to get a clue as to what motivates satisfied biologists: They love what they are doing. More often than they will admit, their love is for the organisms themselves. Such feelings tend to slip out around tables at Friday afternoon watering holes and in seminars presented away from home. An ecologist anxious to maintain his place in the reputation dominance hierarchy talks about his "good system to study the interactions of community members," and drones on about hypotheses, resources, alpha values. Over a beer he says he loves grasshoppers. A young behaviorist among senior scientists expounds economic theory as it applies to foraging strategy. With a pool cue in his hand he talks about sparrows. A comparative physiologist gets funded by proposing studies of molecular exchanges between symbionts and hosts. But he begins a department seminar in awe of *Hydra*. These stories are so commonplace among professional biologists that they hardly seem worth repeating. To a person who's never sustained a focused intellectual effort for as long as a decade, however, they reveal a critical element in the life of a biologist: the fundamental love of chosen beast.

This characteristic can serve as a good guide for the beginner: Study what you personally want to study, not what someone else wants, tells, or needs you to study. Let curiosity and desire dictate your initial decisions. It is all right to study an organism simply because you are fascinated by it. This kind of attraction will sustain interest on those long days in which experiments seem not to work, hypotheses turn out to have flaws, and an incubator breakdown costs you a month's labor.

Furthermore, the body of biological literature suggests this advice is sound. Journals are filled with sophisticated research on various obscure organisms. An hour in the library reveals everything from detailed behavioral analyses of jumping spiders to investigations of protein synthesis in a silkworm oviduct. In a surprising number of cases, these papers contain significant contributions to the understanding of fundamental biological processes. Within a few years, introductory textbooks will use this type of apparently esoteric work to illustrate concepts for people getting their first exposure to biology. A classic case in point is the study of a group of nondescript birds known as the Galápagos finches. Try to visualize Darwin at tea in 1838 explaining to a society matron why he was so excited about these organisms. Hugo De Vries not only rediscovered Mendel's laws around the turn of the last century, but also through his chromosome studies discovered speciation by polyploidy in plants. Put yourself in De Vries' place in the early 1900s talking with the man on the street about evening primroses. Konrad Lorenz won a Nobel Prize for his role in the founding of ethology, but imagine him in the 1930s telling a banker about his first imprinted gosling. In a very real way, you will sense the love for the organisms and the simple desire to know more about them that these biologists must have experienced.

Having made a most difficult decision about what subject to pursue, the logical question follows: Now how do I study it? Instead of launching into a list of techniques and sources of money, I'll again suggest an answer that seems trivial: Budget the time. As a member of a social species, you compete for your own time with several billion other people. They fill your airways with radio and television transmissions, throw their newspapers screaming tragedy and stress on your front porch, take your parking place, make lines and crowds, ask you or tell you to buy this or fight for that, try to seduce you intellectually and perhaps physically, alter the structure of your environment, and, among the most deadly acts of all, adore, laud, and want you if you become famous.

"Budget the time" is, of course, too mild a phrase. What I should have said is take the time, your time, away from the human race. It belongs to you. You don't have an infinite supply; in fact, it may run out tomorrow. At least some of your time must be absolutely inviolate, non-negotiably yours. The serious pursuit of any discipline, not just biology, requires large blocks of time in which you are totally free from potential interruptions, telephone calls, bouncing checks, and the like. Furthermore, your personal time should be your best time, those hours of the day when you know you are most creative and productive. History is paved with scholars who gave the organization their best time for committee work, then tried to fit their science into the cracks. It doesn't work.

There will be readers of this book, maybe amateur biologists, supporting themselves as normal members of society, who will argue that they have to make a living and can't budget their best time for biology. No one can deny that in practical terms there is some truth to such a statement. But even if you consider yourself an amateur, you still owe yourself some non-negotiable time for what you want to study. Society will be better off for your having done this. Society needs the perception, maturity, skepticism, and analytical powers that come from studying something that is highly complex. I suppose these comments could apply to a lot of different fields. They are certainly descriptive of all the people I know whom I would categorize as true, successful, satisfied, productive biologists.

2.

The public envisions the work of scientists as doing experiments. While this view has, in the case of biology, been somewhat diluted by growing ecological (hence outdoor) awareness, it nevertheless remains an aspect of the way you are or will be regarded. Will you do experiments? Yes, as a minimum requirement of your formal training. Certainly molecular biology and all its older relatives rely on experiments, and experimentation is becoming more a part of ecological field research every day. Can skillful experimentation be taught or learned? Maybe. Do you have to be a skilled experimenter to become a biologist? No, but your experimental skills will dictate the kind of biologist you can become.

What, actually, is experimentation? It is the manipulation of conditions in order to reveal or produce observations that contribute to the solutions of puzzles. For the beginning biologist, expecially one who tries to gain exposure to original science, the manner in which biological research is presented may disguise the role experimentation played in a particular project. The problem is this: Formal biology as presented by professional scientists is typically (1) devoid of major mistakes, and (2) quite rational. These properties, however, usually accrue after the work is done. Major mistakes are generally caught by the anonymous reviewers and corrected before a paper is accepted for publication. The rationality is provided by the selection, for the written paper, not of data but of the lines of research that actually seemed to solve a problem.

Furthermore, the manner in which biology papers are constructed is somewhat counter to the actual intellectual processes that occur when the research is planned and performed (see also section 7 below). This discrepancy was pointed out to me recently by a graduate student in her answer to a written doctoral comprehensive question. She had been asked by a faculty member to determine whether "zoology was a science" and to defend her answer. In her responding essay she observed that papers are written according to this structure: introduction, methods, results, and discussion. If they were written in a format more consistent with the scientific method, the headings would be: theoretical basis, hypotheses tested, hypotheses rejected and accepted, theoretical implications. Her contention is indeed true. The way we write about our work often hides the way we really do it.

In practice, even the most successful people make mistakes, do experiments that don't work, construct and try to test faulty hypotheses, have manuscripts and grant proposals rejected, and go back to the lab to correct these problems. Over a period of years, one does eventually solve enough puzzles to build, in retrospect, a fairly logical sequence of thought from a vast supply of information which one then presents to the public. In order to build the supply of information, however, you must either successfully experiment, or carry out its philosophical equivalent, devising (1) testable hypotheses, and (2) the methods for testing them.

Testable hypotheses are often ballyhooed as the fundamental element of the scientific method. Hypotheses and their testing are in fact the basis of the reductionist approach to science, in which one sets about to solve a puzzle (= problem) by reducing it to a series of smaller ones. The puzzle is usually well defined: How is the casein gene regulated; what adaptations allow one species of crab to occupy an estuary while another is relegated to the benthic? The solutions involve posing a series of smaller questions that are answered by use of testable hypotheses—i.e., propositions that can be validated or invalidated. Sample hypotheses for the questions above might be: (1) There is no difference in the amount of casein produced by mouse mammary glands *in vitro* in the presence, as opposed to the absence, of the hormone prolactin, or (2) there is no difference in the survival of estuarine crab Species X and benthic Species Y when both are maintained in twenty percent seawater. The result of one test leads directly to the next logical question. At the end we derive a solution that consists of a series of descriptive statements or conclusions that, when taken together, seem to answer our original question, at least at a level that will satisfy us for the present.

This short discussion also illustrates what Thomas Kuhn calls "normal science" (see *The Structure of Scientific Revolutions*), in that the problems have been chosen within a paradigm that affirms the importance of knowing how the synthesis of casein is regulated or what factors drive speciation in crabs. These restricted puzzles are expected to have theoretical implications—i.e., to shed light on larger puzzles, such as overall genetic regulation in vertebrates or the general phenomenon of adaptation.

When a puzzle is biological, it can be assumed to be at least an order of magnitude more difficult than it is thought to be. A sense of how to solve such puzzles requires a certain background of knowledge and experience. The knowledge you will bring to bear is that gleaned from published papers, classes, seminars, conversations, and original observations. Experience is derived from research. There is no way to gain puzzle-solving experience except to do research, and actually write the hypotheses, make the mistakes, mix the chemicals, scrounge up the equipment, record the observations, and draw the graphs yourself. In the end your brain will contain lessons, observations, and intuitions that are unconsciously organized into a puzzle-solving mechanism. The next puzzle you attempt to solve will not seem so formidable. You may not be any more successful in solving it, but knowing the territory, you will be more patient.

This patience will be tested many times throughout your career. The basic activity of the scientific method, hypothesis writing, is in fact an art. You are the only one who can teach yourself an art. One useful technique in learning to construct hypotheses is to put yourself in a certain frame of mind, allow your knowledge and experience to stew for a while, and allow alternative hypotheses to multiply and compete with one another until you select one to test. At this point your exposure to the arts might pay dividends. The writing of a hypothesis is a creative act as surely as is the writing of a novel or the painting of a picture. Artists always reject more of their own work than they show. The same can be said for scientists and their hypotheses.

3.

Biologists work not only in a largely ignorant nonscientific social environment, but also in the very conservative one of normal and reductionist science. The conservative elements will constantly question your choice of puzzles, thus exerting powerful selective pressure on your career. I have seen this kind of pressure operate again and again on beginning graduate students. Some of the most conservative people in their environments are, in fact, their own peers. It is foolish to underestimate the effects of a question such as "Why are you studying the nonpathogenic parasites of grasshoppers while I'm testing foraging strategy theories?" when all you can answer is "I like them." A senior scientist privy to this conversation might easily be reminded of days when his own children had to answer their junior high classmates' "Why are you wearing that kind of a shirt?"

If you are a creative person, this type of pressure can become very frustrating. Creative people do not like reductionist science. Neither do multitalented, worldly, widely read, deeply philosophical people. Yet biology is so broad a field, with so many interconnections with other fields, and a source of problems, questions and organisms that is so rich, that creative minds are at a premium. Biology simply cannot afford to drive anyone away from an original line of study with pressures such as I've just described.

I mention the above problem because beginning students often judge a field by the image that is presented to them by its professionals. This means that the "important" questions and problems are really ones that the present generation of scientists has been working on for some time. In essence, college professors tell you what puzzles are significant. The vast majority of professional biologists at universities pursue normal science (see Kuhn), which produces papers fast enough to get tenured and granted, the two requisites for keeping a job in academia. History has shown, however, that science progresses by revolution. Paradigms do not evolve, but are replaced by new ones: Einsteinian vs. Newtonian physics, Du Pont vs. phlogiston chemistry. You can see a recent example of revolution at work by comparing cell division as it was presented in an introductory text from the early 1950s with the same subject in the latest books. Thirty-five years ago we concentrated on chromosome movements in mitosis and considered interphase as merely a "resting" stage. Today introductory texts cover mitosis out of tradition and as background to developmental genetics. The important action of cell division, we have learned, occurs during interphase; the various chromosome movements of mitosis have little bearing on the problems of regulation of DNA synthesis. In short, the paradigm that dictates valid puzzles and their solutions in cell division today hardly includes the picture of cell division from 1950.

Because biology is always vulnerable to revolution, the creative person always has an opportunity to start one. There are few activities more exciting than rebellion. It's in the same category as a major upset in college football. Reductionist scientists, like coaches of highly ranked teams, hate it, which makes it all the more fun. Post-revolutionary behavior in science has some similarity to that which occurs after political revolutions. Some scientists are left hopelessly behind and lost, others endorse the new paradigm with inordinate zeal. But it's never smooth sailing, especially for the rebel.

A fantastic example of such an event happened in my own lab, and the memories remain some of my strongest. The principal actors were a highly creative graduate student and a totally disrespectful protozoan parasite, *Herpetomonas megaseliae*. *Herpetomonas megaseliae* is a flagellate protozoan parasite of *Megaselia scalaris*, a phorid fly. Monogenic flagellates (requiring only one host species) that are carried by insects are not supposed to infect vertebrates, although closely related digenic forms (requiring two or more host species) can cause fatal infections in humans. In fact, pre–World War II workers had tried to infect vertebrates with monogenic insect flagellates, mainly out of curiosity, but had failed. Their failure, in effect, established a paradigm that held that studies to test the infectivity of monogenic insect flagellates for vertebrates were invalid puzzles. Indeed, current prevailing thought allowed no context within which to interpret positive results, should they have been obtained. Such results would have been considered an isolated observation with no relevance to the problems of parasite/host cell relationships in disease-causing organisms.

My graduate student challenged the prevailing paradigm (on purpose, I believe!) and achieved his contribution with a set of experiments that were very elementary in conception but required a certain love of the material. The student simply cultured his *H. megaseliae* and fed them to mouse peritoneal exudate cells in flat-bottomed Leighton tubes with cover glasses. He then removed the cover glasses at various times, stained the cells, and studied them. What he discovered was that the protozoa had been taken up by the phagocytic cells and had then changed into a round body form indistinguishable from that of a related species that would infect vertebrates. With this evidence, the student knew what to look for after he injected the insect parasites into living mice, and, needless to say, he did find his round-bodied organisms in the mouse spleen. There still remained the question of whether the parasites were alive. He showed that they were by replacing the original mouse cell culture medium in his Leighton tubes containing the mouse cells with flagellate culture medium and incubating the tubes at 25C. The round intracellular forms left their host cells and multiplied as elongated flagellates. He then went back to the mice and cultured spleen tissue to demonstrate that his parasites could remain viable in the vertebrate for a few days.

What he had done with his experiments was to determine that a noninfective species could carry out the initial morphological changes and early survival typical of an infective species, but could not multiply within mammalian cells. Since infectivity demanded not only morphological change but also long-term survival and multiplication of the parasite, he had discovered the specific point of difference between two related forms, one infective to humans and the other supposedly not. This work had major evolutionary and medical implications. The differences that distinguish infective from noninfective parasites can be considered evolutionary accomplishments that allowed this group of protozoa to colonize vertebrates. The fact that such colonization occurred at some time in the past is a medical problem in the tropics today. The experimental results suggested, for example, that drug treatment research might profitably focus on the specific parasite functions that allow intracellular survival beyond 12–24 hours.

The student's manuscript, well prepared, and extensively documented with a series of photographs, was rejected by the *Journal of Protozoology*. A reviewer called him a liar, said his photographs were not what the student knew them to be. (I can still quote the review: "Fig. 2 cannot be a picture of *H*.

megaseliae.") After much ranting and raving, the student repeated the work but this time prepared his infected cells for transmission electron microscopy. He submitted electron micrographs with a shorter manuscript to another journal, to which it was accepted instantly.

This case is worthy of some analysis, for it demonstrates a number of things about the practice of biology. First, reviewers often approach a manuscript with preconceived notions because the history of investigation dictates what they expect to find in a paper. To a great extent science must operate in this way. Consider the chaos that would result if published work did not influence reviewers. Second, the case demonstrates that biologists at the beginning of their careers are able to make major contributions. Third, the fact that results have been published doesn't mean the work is beyond reexamination. The intellectual climate in which observations are made continually changes. As with mutant alleles, ideas can take on new values in a changing environment. Fourth, technology validates observations in the minds of scientists. The TEMs showed exactly the same phenomenon that the light micrographs had shown, but the characteristic image of an electron micrograph validated the work. Biologists believe—rightly, in most cases—that the electron microscope reveals truths that cannot be seen at the light level. In this case they simply believed what they had been shown earlier but had not been convinced by. Finally, although the paper has been cited many times, the contribution has yet to influence the context in which this line of work is done. Other authors have referred to the paper's description of changes in parasite microtubule arrangement that follow phagocytosis, but they have ignored the fact that the experimental results have evolutionary implications. Someday another student will come along and establish the new context by "seeing" in the work something that senior workers, restricted by their own paradigm, cannot see.

What does this anecdote mean for one who would be a

biologist? It suggest that research need not involve the latest high-tech problems and equipment to be of significance, that the use of an organism's genetically determined capabilities is in itself a technique worthy of acquisition, that you will have to present your work in a manner established scientists can understand in order to get it recognized, and that the leaders will not always use the research work in the way you know it to be most useful. This student's synthesis was original, although that originality involved a new combination of old ideas, established techniques, and confidence that the organisms involved possessed certain stable biological characteristics. Most importantly for the aspiring biologist, the case bears out, as could literally tens of thousands of others, the fact that it is the individual who is ultimately responsible for the synthesis, not the group. The group—i.e., the paradigm—is conservative. This virtually immutable rule of intellectual endeavor is one of the most difficult lessons for a young biologist to learn.

4.

In the anecdote I just related, a major role was played by the organism itself. It is important to recognize this role because there is a subtle but sophisticated quality to research that utilizes the unique characteristics of a species as an integral part of the work. Biologists who understand how species' traits can help answer their questions seem to be more in control of their material than those who do not. Organisms will always do things for you that you cannot do yourself. They will predictably carry out certain reactions, regulate internal environments, respond to experimental conditions, and enable you to make observations to compare with previous or concurrent ones. This power is implied in their taxonomic names. Unless proven otherwise, an organism's name, including its strain or stock designations, guarantees a specific set of genetically determined traits. The assumption rests on the species concept. If we can't assume a species has a history of adaptation to certain conditions and possesses a defined spectrum of functions related to that adaptation, then we are out of business.

The love of chosen organism is what will give you the touch, the patience to culture a wild bacterium, virus, plant, or animal, the intuitive sense of how to design studies that make the most of unique attributes. If you love an organism, you will also recognize the difference between it and related forms, which in turn will make you aware of the potential use of other organisms. This awareness leads to the comparative approach to biology, which has yielded major concepts in our field since the time of Aristotle. For example, in the 1920s Frederick Griffith's studies on *Streptococcus pneumoniae* would not have yielded the concept of bacterial transformation had he not compared an infective (S) with a noninfective (R) strain of bacteria. The S strain did something the R strain could not do (kill mice), and Griffith ended up discovering a method for transferring the unique property of one strain to the other. Similarly, François Jacob and Elie Wollman's classic work on the structure of bacterial chromosomes was clearly dependent on the study of several strains of *Escherichia coli*, each of which transferred its genes to a recipient conjugant in a unique and predictable sequence.

The comparative approach is not characteristic of a certain fraction of scientists who parade as biologists but present papers, at national meetings, in which they ignore the source of their material. These are people who can analyze the results of restriction endonuclease digestion of mitochondrial DNA but can't tell you the family characters of the species from which the mitochondria came! Most likely another scientist, one whose perspective is broad, and who understands the importance of asking how organisms differ, will use these kinds of isolated observations, putting them together with others, to build a general concept.

This is my recommendation for the serious student: No matter what your area of interest, be it genetic engineering, biochemistry, molecular genetics, or ecology, take the time to study whole organisms. Take as many courses as you can in taxonomy, nomenclature, advanced organismic biology. Take them with an open mind; learn to love something. Go to a field station. Pay your dues classifying invertebrates or vascular plants. In the end this experience will give you the comparative insights that separate true biologists, with something significant to say about life on Earth, from the equipment jockeys.

The ability to manipulate living material is one of the most valuable skills a biologist can acquire. By "manipulate" I mean capture, prepare, maintain, culture, and most important, place in a certain functional or structural state or relationship. This general set of activities is often assumed to be summarized in the word "standardize." But that term is not inclusive enough. Certainly you will want to be able to choose material, from the genetic background of mice to the cycle phase of cells, that is within a known set of conditions for repeated experiments— i.e., standardized. But a manipulator can also choose the condition and relationship that he or she wishes to standardize. Often such choices take one well beyond published methods or instructions. As with the comparative approach, manipulative ability will separate you from the rote biologists, provide you with biological materials others do not have access to, and allow you to exploit organisms' genetically endowed traits. If you don't learn how to manipulate living material you may have trouble your entire career. If you are not willing to do it, you will eventually have to find the money to hire someone who is, especially if you expect to work seriously in any area of "modern" biology. Then you may end up as the distinguished scientist whose reputation depends on a technician.

The best training in manipulation is normally found in genetics and microbiology labs. There you will discover that this phase of biology is a cooperative venture something akin to agriculture in that you must cooperate with nature in order to produce the desired results. (The two other great cooperative ventures are medicine and teaching—see Adler, *The Paideia Proposal.*) What you want to become is a successful farmer of fruit flies, corn seedlings, *Neurospora,* or various bacteria. A good farmer anticipates, thinks like his or her organisms, empathizes with them, is sensitive to their needs, derives enormous pleasure from their growth, and harvests them. An off-color batch of culture medium rings an alarm bell. A farmer checks an incubator thermostat by reflex action whenever he or she enters a lab. The ultimate value of this approach is that it will be transferable to organisms and cells the farmer has never seen before.

An exceptional demonstration of manipulative skills took place in my own laboratory when a post-doctoral researcher, with a dissertation on the reproductive biology of crustaceans, came to work on protozoa. I felt his skills with the electron microscope would allow us to address some unusual questions about intracellular parasites. But after a few weeks, it became obvious that he had come to learn something new, not merely to perform for us the same kind of work he'd done for the previous three years. I still had no idea that I'd recruited one of the most incredible minds in the world, not to mention a person whose individualism defied description. Needless to say, his EM work was done in a rather perfunctory manner.

In the meantime, he became intrigued by the behavior of protozoan parasites in mammalian cell cultures. Mornings I would often discover evidence that he'd been at work all night. His attention, it seems, had been captured by the published observations of that previously mentioned rebellious student—namely, that some parasite species survived longer than others in cultured mammalian cells. What mechanisms operate, he asked, to produce this differential survival? What was involved here was the distinguishing feature of an infective, as opposed to a noninfective, parasite, a distinction that should have obvious medical importance.

In retrospect, the experiments he designed to answer his question seem to have been inspired by one of those flashes of insight that come only a few times in a person's life, if at all. He began by reducing his original question to a more specific, and answerable, one: Can an infective parasite alter a mammalian phagocytic host cell in some way that promotes the survival of a noninfective parasite? The noninfective organisms were to be his probe—that is, a target whose survival would indicate altered host-cell function (but would not necessarily reveal the specific nature of the alteration). The host cells he selected, peritoneal exudate macrophages, were cells that could be expected to destroy parasites. Indeed, his preliminary work showed that a parasite species that could not infect mice was also rapidly digested by these cells in culture, while a species that was infective for mice survived significantly longer. The question he now posed, however, implied the infective parasite, instead of simply passively resisting host cell digestive enzymes, might "actively" contribute to its own survival. All of the biological materials to answer this question were at his disposal. The laboratory procedures he would perform, though, required unheard-of manipulative skills.

This young scientist envisioned a set of experiments in which he would infect mammalian cells with one parasite species, then follow that infection with one using a second parasite species. The first species was to be an infective, the second a noninfective, one. In order to accomplish this, he had to standardize both his mammalian cell and his parasite cultures, making sure all were of specified ages and cell concentrations, and in the proper incubation medium. He also had to standardize his protocols—i.e., times of infection, relative numbers of parasites and host cells, and scheduled changes of culture medium —so as to produce predictable results. That is, he had to know that if he gave the host cells so many parasites, then he could count on so many surviving (or not surviving in the case of the noninfective species) after twelve, twenty-four, and thirty-six hours. He had to repeat these standardizations using parasites killed by various means, to show that a live infective parasite could alter the host cell while a dead one could not. He had to perform identical standardized infections with a variety of other materials to demonstrate that the host cell alterations he observed were specific to the parasites he was studying. He carried all this work out under sterile conditions, without the use of antibiotics, which he felt might interfere with biological events. His list of scientific equipment consisted of a balance, centrifuge, incubator, and hood.

His first publication demonstrated a principle that the scientific community, caught up in immunology, had not explored: An infective parasite is successful because it alters its host cell to make it a suitable home by inhibiting the cell's ability to digest the parasite. The noninfective parasite does not have this power. Over the better part of three years, the post-doc had manipulated hamsters, their peritoneal exudate cells, two different parasite species in culture, and combinations of cells and parasites to produce every few days a set of experiments in which the variation due to laboratory technique was kept to a minimum. He standardized ages of hamsters, times at which all manipulations were carried out, the ages and metabolic states of his parasites, and the culture conditions of his mammalian cells. In the end, the careful control he established with his manipulative skills allowed him to combine cells in a way that used the cell's known properties as an integral part of the experimental design. No analytical machinery, no matter how powerful, can replace this kind of human talent.

What has happened to this particular talent recently? His second paper, from these experiments, demonstrated that the infective parasites were liberating materials in culture that the noninfective parasites were not, and that these materials could, by themselves, inhibit host cell digestive function so as to allow survival of noninfective parasites as well. He had thus begun to decipher the molecular communications by which the parasites altered their host cells. One anonymous reviewer questioned his statistical methods. In response, the post-doc began a serious study of overdispersed (variance greater than the mean) distributions. The manipulative power of the computer, combined with his newly developed mathematical skills, were too seductive for such a mind. He is now head statistical consultant at the computer center of a large state university.

5.

Research always goes more slowly than people think it should. An elementary exposure to statistics reveals the reasons why. If observations are normally distributed, ten or more repetitions are generally needed to satisfactorily estimate means and variances. Much published experimental biology has attempted to evade this stricture through claims of analytical precision. The worn computer rule (garbage in, garbage out) probably applies here; phrased biologically it becomes: So your measurement is of great accuracy and precision; now what exactly is it that you've measured? The question leads us back to the choice of material to study and the manipulation of it to achieve desired results.

Few better examples exist in the history of science, of high tech superimposed on organismic naivete than the case of the HeLa cells (see *Science 81*, April, 1981; "The Immortal Cells of Henrietta Lacks"). HeLa cells were first isolated at Johns Hopkins University in 1951 from a cervical tumor of a woman named Henrietta Lacks. The cells could be grown very easily —too easily, it turned out. In 1973, Walter Nelson-Rees from the University of California-Berkeley cell bank determined from unique chromosome structure and enzymes that a number of cultured cell lines were not what they were purported to be, but instead were HeLa. It is almost inconceivable that a single cell line could escape from petri dishes to contaminate cell cultures in lab after lab. Yet that is apparently what happened because of sloppy lab technique. In the late 1970s, untold thousands of dollars' worth of cell biology research had to be reinterpreted because investigators who thought they were working on various cell lines were actually working on HeLa. In retrospect the blame for this situation must lie with an approach to biology that subordinates the biological material to the processes studied and the analytical techniques employed.

Research also proceeds more slowly than expected simply because everyone makes mistakes. True scientists, and most particularly true biologists, budget for mistakes. Furthermore, those who've read their scientific history are always alert to the possibility that an error may reveal secrets that would have remained hidden had things been done "correctly." For several years a letter to the editor, clipped out of *Science*, remained taped to my incubator as a constant reminder of this chance. The letter described the author's student experiences in a bacteriology course in the 1920s. His laboratory instructors were constantly reminding him that bacterial plates with mold growing on them were ruined and should be discarded. Years later, Sir Alexander Fleming noticed that bacteria were lysed at the edges of mold colonies on contaminated plates. He grew the mold separately and discovered the used mold culture medium could prevent bacterial growth. The active compound was, of course, penicillin. In the course of a move to a new building, this letter was lost. For the purpose of writing this paragraph, I asked a colleague who always saved such choice literary contributions if he could tell me the author of that letter. He couldn't; it seems his copy was on his bulletin board amidst dozens of other items such as cartoons, news stories, and interesting pictures, and somebody had stolen it. Evidently, that single letter published long ago still had some meaning for a person wandering the halls of a biological science building.

In addition, biologists must allow for variability within the organisms they are studying. The variability of living material is, in fact, its evolutionary strength. In general, the greater the variability in a phenomenon, the greater the number of observations required to validate or invalidate a hypothesis. A single species may be extremely variable in a multitude of functions or structures. In some cases—e.g., with antigenic variation in flagellate protozoan parasites—the variability is itself an adaptation, and biologists study it as such.

Variability is anathema to a reductionist mind, a source of intrigue to a creative one. It is for this reason that scientific progress ultimately depends on both kinds of personalities. Thus in narrowing a puzzle into its solvable components, the normal scientist seeks comfort in the familiar, the predictable, and that which can be reliably standardized. The creative mind, bored by such prospects, searches chaos for the answers to grand questions. If history is any guide to the future, so long as creative minds explore chaos, there will be work for reductionists. And, so long as normal science reduces puzzles to their solvable components, our detailed knowledge of nature will increase on a solid foundation.

In watching the rate at which the careers of my fellow scientists develop, thrive, and wane, I have often wondered: Do the historical processes I've just mentioned, the interplay between normal science and revolution, operate at the individual level? Yes; but we rarely think of an afternoon's work in historical terms. On the second floor of the University of Nebraska library there is an art gallery. For two months, the paintings of Francis Lee Jaques hung there for all to see. But the place was deserted. No one seems to remember Jaques' contribution to our view of nature. On a whim, I asked my lab instructors to take their introductory biology students to see the paintings. After all, Jaques documented a Central Plains that we have transformed by plowing to the fence.

But my whim violates departmental desires; there are repercussions. I'm told my students will be responsible for the missed lab material on the departmental final. It doesn't matter; Jaques hangs for two months, one can study osmosis any time. One of my teaching assistants compounds the problem, however. The lab originally scheduled for this week concerns diffusion, osmosis, and active transport, concepts that few business majors consider critical to their financial success. The exercise is also among our most boring; nonmajors remember it largely as "watching dye go up a carrot." So the teaching assistant makes this assignment: Critique the Jaques paintings in terms of this week's exercise—i.e., discuss how active transport, diffusion, and osmosis contribute to and control the circumstances depicted in the paintings. His lab section visits the gallery for an hour. By the time he gets back to his office, the fur is flying. Sarcasm walks the halls. He's called "cosmic," and compared to an unruly faculty member. The incident, he is told, will have a negative effect on his career. I look over some of the answers his students have given. For the first time in my memory, business majors have put osmosis, active transport, and diffusion into some kind of a meaningful context.

The experience just described raises a difficult question for scientists: Do I repeat an experiment today, or simply watch the flowers and think some strange and controversial thoughts about the organisms I love? The most successful biologists do both, easily and naturally, without guilt or regret, although it slows their "work." They remain happy, challenged, self-contained, and confident. Their publication records and grant dollar totals lie somewhere between those of the reductionists (high) and the creators (low).

6.

The practice of biology, like that of any science, requires money. Funding can come from a variety of sources: government (at all levels), industry, private foundations, personal inheritance. Molecular biology, biochemistry, and cell biology in particular can all be very expensive, so if you expect to practice these kinds of biology then expect to take on the task of raising tens of thousands of dollars annually for at least a decade. Field biology and ecology, while often cheaper in terms of capital outlay, are nevertheless costly in other ways— for example, travel, assistance, personal time, and inconvenience. On the average, a professional biologist practicing in the 1980s needs at least five thousand dollars a year for research in order to publish regularly and develop a career of serious study. If his or her research area is cell or molecular biology, that figure must be increased by an order of magnitude. An amateur needs less, but must make up for the lack of financial support with a storehouse of ideas. In this regard biology resembles some other human endeavors in that ingenuity is the equivalent of material wealth.

Although this book is not intended to be a guide to successful grantsmanship, as true a picture as possible of the economic side of biology is necessary for anyone contemplating entering the profession. Basically, the situation is this: To be a biologist you must be willing and able to raise or make money to support your work. In this regard, science is similar to art, writing, and film making and different from other occupations in which your work makes money to support the trophic portions of your life. "Work" to a biologist is biology, but to the outsider may take on the appearance of play, or activity that has a sense of childishness, such as seining fish or watching birds. Biologists themselves tend to not take other kinds of work very seriously except to the extent that they provide funds for biology. This apparent mixup between work and play results, of course, from a biologist's values.

Raising money for anything requires (1) effective communication and (2) a willingness to bend your ideals at least a little. By "communication" I mean successful writing. In preparing a grant proposal you must be able to build an easily readable document that makes a convincing case for your project. You must be able to force a reader to follow your train of thought, and you must progressively impart so strong an air of significance about your work that the reader will feel confident you are worthy of an investment. The necessary writing skills are close to those of the novelist and historian. Although they are neither taught nor learned in science classes, they are ultimately required to do science at the highest levels. They are one of the best reasons for taking your humanities seriously.

The need to develop communication skills makes it imperative that you try to publish as early in your career as possible. You may find yourself in an atmosphere that warns, "Don't try to publish until you have something truly significant to say and have solved a problem completely." This ideal is largely hogwash. The history of science, not individuals, judges what is significant among published works, and for an individual to preempt this function is an act of grand arrogance. The anonymous reviewers of journal manuscripts will try it, of course, because they've been asked to by the editor. Furthermore you're probably never going to solve any problem completely. Your sense of the state of knowledge in your special area of interest should guide your decision to try to publish. If your observations seem to fall somewhere in the range of significance of already published material, and seem to add something new to our knowledge, then they are probably worthy of at least a short paper. If the anonymous reviewers for a journal decide the material should not be published, you don't have to accept their decision gracefully—but there is no way you can force your material into a journal. If, however, some professor argues against publication because the literature is already too cluttered, be suspicious of the opinion. In the long run it is more important for science that its young people learn to communicate than for the literature to be uncluttered. The journal reviewers are of great help in acquiring writing skills. And you will never feel the full impact of the anonymous peer review system in science until you try to publish. The sooner this happens the better off you, and science, will be.

For a scientist, publication represents many things: an accomplishment, a summary of an often long and challenging project, the culmination of a good idea. But most important, it is also proof that the scientist can actually do something. This proof is usually necessary before you can hope to raise thousands of dollars a year for several years. Furthermore, publication is proof that a person can think, convert ideas into reality, and influence the scientific community. A significant section of every grant proposal is the one that cites the record of an investigator's past accomplishments. One of the major reasons why some great ideas are rejected by funding agencies is that these ideas seem to extend beyond a researcher's proven (by publication) abilities.

I also mentioned that a willingness to bend your ideals somewhat was also a prerequisite for obtaining funds. It is not my intent to be cynical in this regard, I simply want to remind you that dollars rarely come without at least a few strings attached. In the matter of ideal bending, I have seen the extremes among my colleagues. There are those for whom the grant dollar is equivalent to success and whose research programs seem to evolve in the direction of the greatest supply of money. Most of these people would deny that they are bending their ideals. They might even claim, sometimes rightfully so, that they can "bootleg" funds supplied for well-defined, often quite applied projects and sponsor truly independent research. On the other hand there are those who are determined to pursue a line of research regardless of its funding potential. In both cases, the grants, or lack thereof, don't seem to particularly affect the quality of the science, but simply the areas in which the contributions are made and the rate at which they are accomplished.

7.

Publications not only serve as a means for developing writing skills, they are also used by scientists to evaluate other scientists. The exact manner in which they are so used, however, is not always clear. It is common to assume that for those who publish easily, things go smoothly, salary raises are generous, and reputations and respect grow steadily. Experience, however, does not support this impression. After a certain number of years, publication may be seen as an unfortunate but also necessary and tolerated part of scholar performance evaluation systems. In other cases, the submission of a manuscript becomes a highly personal act in which a researcher thrusts his or her contribution at the field, daring the journal to accept it. On the average, relatively few, among those who call themselves scientists, regularly publish serious work for their entire careers.

Some people publish early and easily, others rarely and even then with great difficulty. No one knows what combination of inherited traits, childhood memories, student experiences, and self-expectations contribute to the two extremes. I have known people at both ends of the spectrum. One of my former students, for who knows what reason, once sewed together a lab coat made of fake lizard skin. He also spent long nights at research, sprinkled his conversation with creative profanity, made a "D" in biochemistry, and generally worked his way into a state of persona non grata in our department. He is now one of our university's most successful graduates, holding a full-time research/curatorial position and administering thousands of dollars in grant funds. He published a great deal as a student and continues to do so. Another of my colleagues also publishes very frequently, administers thousands of dollars in grants as well as a lab full of post-docs, and is invited to be a speaker and panelist all over the world. He has stopped attending faculty meetings. I sat one day in a chairman's office and heard his work discussed: Have you really read all those papers? There were, in the questions, implications of emptiness. I remember spending, shortly thereafter, some long hours just thinking about the significance of publication in the career of a scientist.

Another colleague, an invertebrate zoologist, died recently. His wife had passed away a year earlier. He had been a soft-spoken, retiring person, and had neither published nor written a grant proposal. He had, however, won the university's Distinguished Teaching Award, one of the highest honors our institution can bestow upon a faculty member. His funeral was sparsely attended, and featured taped music, a pauper's eulogy, and a reading of "Thanatopsis" by a minister who'd never known any member of the family. I sat in a narrow pew feeling like a bit player in some western movie in which a few lonely people gathered in black on the windy top of some forgotten Boot Hill.

Later, after weeks of wrestling with the decision, I walked into my chairman's office and volunteered to teach my departed colleague's course: junior-level Invertebrate Zoology. The subject area is limitless, as are the possible approaches; the history of invertebrate zoology reads like an epic struggle between competing grand ideas. If here on the prairies roundworms alone can assault our world view, what untold power must lie with the marine annelids. The challenge was accepted. After an appropriate grace period, I entered his lab to begin sorting through the resources at my disposal. It was almost an archaeological experience. One can indeed reconstruct a life, a pattern of thought, an underlying philosophy, a set of values, from the things a person leaves behind. This particular lab was a museum and library combined. The trail of my colleague's intellect led through a bewildering complexity of specimens, but always with surefooted underlying principles: variations on a theme, filter feeding as a metaphor, frail survival in harsh environments. Amidst the shells, the series of crabs, I thought: It is easy to equate a long bibliography with quality, a short one with the opposite. But these equations are the stuff of administrators. They have little to do with real intellectual worth, everything to do with advancement up a system. Respect, and the dignity that accompanies it, should not necessarily be tied to one's bibliography.

Publications are also the major vehicle by which the scientific community establishes its own concept of the state of the art. They come in many forms, from abstracts to multiauthored series of books. The volume of published material in biology is absolutely enormous. This bulk will force you to specialize, if you are to keep abreast of the developments in any one area, and in the process you will continue to be haunted by good ideas you wish you had time to pursue. A multitude of services help you deal with the literature: *Current Contents*, computer search services, abstract journals. There is so much written material produced around the world that you as an individual can assume no responsibility for the control of its growth. Instead, if you're going to be a serious biologist, you will have to add to it.

The mystery of publication disappears quickly with experience. It all originates with some human sitting down at a typewriter, word processor, dictaphone, or, as I am doing at this moment, with a pencil and a piece of recycled computer printout in a corner of the Student Union coffee shop, and putting words to paper, tape, or disc. Writing, even scientific writing, is a highly individual act. You must choose which words go in which sequence, which graphs to present, and which references to use. As a guide you will generally have these helpers: (1) the "instructions to authors" in your target journal, (2) the structure of published papers in your immediate area of interest, and (3) the inside advice of people who've done it.

The first two of the above guidelines are fairly straightforward. Instructions to authors are designed for the benefit of editors, typesetters, and layout people, not the preliminary readers. Anonymous reviewers are usually people who've published, and so are accustomed to the format of submitted papers and able to adjust their reading styles accordingly. If you don't follow the instructions, most editors will simply return your paper unread, so that on your second attempt you will be alert to the differences between manuscripts submitted for journal publication and those submitted in freshman English.

The structure of published papers in your area of interest is not always easy to discern. You must read the original literature for structure separately from your reading for content to get a feel for the economy of style, sentence structure, paragraph lengths, and use of certain pronouns typical of the work in your field. Ask yourself: How has the author actually built this paper? As in the case of the young novelist struggling with self-doubt, one of the best exercises you can do is type a published journal article on your own typewriter according to the instructions to authors. Things have a way of appearing possible after all when presented in that familiar typeface.

As for the inside advice, here is one version. Many of the most successful scientists gather all their graphs and tables, complete, ready for submission, before writing a word, then sequence these items according to the line of thought they wish to impart to the reader. They write the results section first, then the methods, discussion, introduction, bibliography, and abstract, in that order. This sequence of writing helps assure that the only material included is that pertaining to the observations the author wishes to report. More important, however, if one already knows the results and their interpretation (discussion), then it's relatively easy to write a hypothesis in the introduction that makes a paper read like a logical piece of work.

If I thought the foregoing sounded too jaundiced, I would not pass along the next bit of inside information: No matter how logical a paper's train of thought, or how insightful a discussion, the most knowledgeable readers will usually read the results first, then check the methods to see if they support the observations, and then apply their own interpretations to the findings. Original scientific writing, and the reading of it, are obviously, and often beautifully, utilitarian.

8.

The room is dark; an audience, scientists by their clothes, sits on hotel conference room chairs. A slide is projected. Filling the screen is a table in which five years of research and fifty thousand grant dollars are summarized. There are groans, audible and embarrassing. The numbers cannot be read even from the front row. The speaker of international repute apologizes for his slides. One more time I wonder: If he knew they were so bad, why did he inflict them on us? It is an interesting question. Everybody knows that in today's world, graphic representation is as basic as The Three Rs. Surely that fact is not lost on top scholars, or is it? Is it beneath their dignity to draw a picture that tells the same thing as the table? Would a picture be as equal to a thousand numbers as to a thousand words?

Respectively, the answers to the last three questions are: (1) Sometimes I wonder, (2) no, and (3) yes. Earlier in this book I discussed the contribution of art to a biologist's training. In this section, however, I am dealing with the practical application of those previous, somewhat theoretical ideas about the role of art in the naturalist tradition. Before, I raised the issue of the relationship between artistic expression and the ability of scientists to convey their thoughts. Now I'm considering the rather mundane problem of drawing a picture.

The artistic abilities of some of our intellectual ancestors call into question our sense of progress. Christian Gottfried Ehrenberg published, in 1838, a two-volume set (text and plates) entitled *Die Infusionsthierchen als Volkommene Organismen (The Infusoria as Complete Organisms)*. He had none of the modern lenses or special lighting equipment we take for granted today. Yet several of his taxonomic descriptions of protozoa still stand as valid. His volumes are in the rare-book vaults, their plates the kind of things cut out with razor blades and sold separately as wall art. To put his work in the hands of a present-day student, seated at the microscope, is to witness a total realignment of mental processes. The Ehrenberg pictures silence all arguments about the relevance of drawing assignments in a biology lab, so that it feels obscene to wonder why we aren't cloning a gene instead. In the presence of Ehrenberg's plates, we are also aware of his respect for an audience. We can sense his thoughts: This material cannot be told, it must be shown; I must take the time and care to show it correctly, properly; I am producing art for the ages.

There is no great mystery to graphic design. But like writing, it comes easily and naturally to some people while others struggle or don't even attempt it. Poor graphics are an insult to an audience, any audience. Adequate graphics involve using a minimum number of lines, high contrast, and conveying less information than most people would like to transmit. Beyond that, graphic communication follows some simple rules: (1) Verticals and horizontals convey stability, diagonals are action; (2) a graph with more than three data lines is disconcerting; (3) a picture of a plant or animal does wonders for a set of data slides; (4) most people trust you've done the basics correctly, so skip pictures of techniques unless they are uniquely integral to your results; and (5) in print, interactions between graphics and text must be easily accomplished in both directions.

The point that seems to have been lost on our scholar of international repute, now apologizing for his second screenful of numbers, is this: No matter how sophisticated the science behind the numbers, once a person decides to present his work visually, that person enters the domain of the artist. LeShan and Margenau *(Einstein's Space and Van Gogh's Sky)* address this problem elegantly. Their fundamental argument is that art relies on feelings, on mental events than can be recognized but not quantified. Such events are all too often an embarrassment to the person who wears his or her science as a badge. And I wonder, as our scholar apologizes for his third unreadable slide to graduate students who've paid two hundred dollars each just to register for this international congress, whether this man was ever told as a child that it was all right to draw a picture?

LeShan and Margenau list four realms of art, all of which are of significance to the biologist wishing to convey information with graphics. The realms are: (1) The intent of the artist, (2) the responses of the beholders, (3) the domain of man-made things, and (4) the domain of the medium. In our apologizing scholar's case, is it his intent to show unreadable numbers, or to convey an idea? He has accomplished the former quite admirably. When he prepared his slides of data tables, was it his intent that the response of the beholders would be to go to the restroom? The domain of man-made things, which includes all objects designed and constructed by humans, might well be limitless. When this man set out to make something, did it have to be a screen-filling table of numbers? Perhaps when we consider the domain of the medium we begin to get an explanation for our scholar's behavior. Late at night, sitting before the fire with a scientific journal, one can leisurely dissect a table, one can choose which numbers to first address. But our scholar of international repute has simply ignored the fact that the printed journal and the stage are two different media.

My scholar, of course, is hypothetical, but his prototype can be found throughout the profession. It will be a shock to you the first time you are told your graphics are unacceptable. At a time when I was selling personal watercolors on the open market, a graduate student asked me to do a half-tone plate for a protozoan species description. I was flattered, and somewhat seduced, by the idea of a traditional plate in a modern journal. I took particular care with the work, arranging the organisms, shading the cytoplasm in ways I had seen so many times on stained smears. The editor, however, in accepting the paper for publication, called my graphics "particularly offensive." He was a man who lived for precision and a certain cleanliness of lines and dots. I redid the figures in ink and hung the halftones on my office wall.

These kinds of cautionary tales will not make you a graphic artist. They will, instead, let you know that you are not alone in your struggle with visual presentation. The most effective advice takes a distinctly artistic flavor: Try it different ways until it works. How will you know when it works? When it feels like it, and when the audience, given the medium, responds with interest. And finally, simply admit that graphic representation is one of those things scientists must do but which cannot, in the final analysis, be quantified and brought to heel with standard methods.

9.

The practice of any discipline involves failure as well as success. Both play significant roles in the shaping of careers. Most people with long and productive professional careers contend with success and failure by defining each in whatever terms seem to work for them. Failure is not always detrimental; success is not always helpful. Of the two, failure, at least of a manageable kind, is likely to bring out the most creative and sustained efforts to solve biological problems in perhaps previously untried ways. Success, on the other hand, often leads to positions where it is difficult to be biologist—e.g., a college presidency. Yet during the early years of a life's work, failure is usually considered bad and success good.

Failure is not really failure if it causes you to seek actively the kind of life for which you are, by temperament or ability, best suited. The aptitude-testing field has an interesting way of viewing professions. It maintains that people are happiest longest in activities that continually test the largest number of their basic aptitudes. Conversely, tasks that require aptitudes which a person does not possess will be a constant source of frustration. And, work that does not test aptitudes will soon become boring. While all these conclusions seem trivially straightforward, the triviality disappears when we ask what aptitudes really are.

Aptitudes are, in essence, physical and mental abilities. They are known to vary among humans. Some "mental" abilities may, in fact, be physical ones, or at least ones imparted by physical traits such as sense organ or nervous system structure. Measurable aptitudes include the abilities to solve a wiggly block puzzle, remember long lists of numbers, distinguish between tones of slightly different pitch, remember random sets of objects, write on paper, decide if two numbers separated by a space on the paper are identical. These tasks are ones almost any person can do. However, when tests are run against time, individual differences become apparent.

Aptitude-testing experts often claim each profession is characterized by an almost continuous use of some aptitudes and not others. For example, photographers usually have high aptitude for distinguishing tones of slightly different pitch, engineers score well on the wiggly block puzzle, and taxonomists score highest on number memory tests. Various careers use and challenge different combinations of these aptitudes. The simple act of matching your personal aptitudes with a complementary career would seem an easy one to accomplish. Unfortunately, the correlations between careers and the kinds of aptitudes being discussed are not general knowledge, especially among high school counselors. Moreover, people, especially young people, often make career decisions based on money, security, parental wishes, and peer judgment, none of which, history shows, have anything to do with specific abilities.

My own experience includes a haunting encounter with a man named Johnson O'Connor and his testing firm, Human Engineering Laboratories. I had been discharged from the army in March and was occupying time, before entering graduate school to "become an ornithologist," by working as a mechanic in a large boat dealership. My father, not overly impressed with my future at the marina, was, I suspect, ready to satisfy himself that ornithologist was a legitimate career for one of his children. So he took me to O'Connor, under whose supervision I spent two full days doing puzzles against a stopwatch. The second evening my father and I sat in O'Connor's book-lined office for a report on the results. He proceeded to tell us that I was not suited to any of the professional biologist activities that I looked forward to with such great anticipation. Most of all, he said, I should not be a teacher. After a couple of hours of hearing him tell me I could be this or that, but really wasn't completely suited for any of them, he said, "What you need is a life of unsolvable problems. You need to be head of some museum." He suggested one, which I'll leave unnamed.

I've spent most of the years since that evening as a teacher. Every class has been exciting and some of them have been downright thrilling. In this case, O'Connor was wrong not about my aptitudes, but about what the profession demanded of them, which is a great deal. Those days in the Human Engineering Laboratories came back to me as clearly as if it happened yesterday, however, when the Chancellor of my university called to ask if I would serve in an interim capacity as Director of the University of Nebraska State Museum, an assignment which I agreed to accept. In this particular case, Johnson O'Connor was right on target. The museum was everything he said I needed! Now as I look back over a quarter century of work as a professional biologist, the only sorrow I feel is that there is not time to study everything. All of the aptitudes that I was told had to be exercised have in fact been used, some almost hourly.

Biology is an exceptionally varied field which provides for literally hundreds of kinds of biologists, each using a different set of aptitudes regularly. The physical and mental skills which are used by an ecologist studying the breeding biology of terns may be totally different from those used by someone doing research on the regulation of mitochondrial DNA synthesis. Questions asked by taxonomists may in themselves reveal aptitudes not used by an electron microscopist. Biology requires this breadth of interest. Progress in our understanding of ultrastructural anatomy will eventually come to a halt if taxonomists are not here to tell us what we are studying. New ways of classifying organisms, revealing important relationships between groups we once thought we understood, will never be introduced if biochemists, mathematicians, and philosophers don't continue to ask questions about the fundamental differences between biological materials.

Success and failure, along with interest and pleasure, are probably your best guides to the kind of biologist you can and should try to be. It may not finally be necessary to know your specific aptitudes in order to match them with a branch of biology. After all, you do have senses of satisfaction and accomplishment. When these senses are stimulated by the study of certain biological materials, then you are probably investing your time in a manner most consistent with your talents. On the other hand, if your studies seem to produce only a sense of failure and frustration, then it may be time to ask the honest question: Am I really doing what I do best, what suits my abilities? It seems, after pages of discussion of the practice of biology, we have come full circle, back to the hardest decision of all: What should I study?

3.

TEACHING

AND LEARNING

> . . . almost all great scientists—
> those who learn to cultivate insight—
> learn also to respect its mysterious
> workings. It is here that rationality
> has its limits.
>
> —BARBARA McCLINTOCK

1.

There is a myth afoot among the learned that holds that there really is a distinction between teaching and research. Or put more forcefully, some academics believe there exists a true separation between that activity intended to expand our knowledge (research) and that activity intended to perpetuate our ability to expand our knowledge (teaching). This issue needs to be addressed because the public has also accepted the notion that the two activities are separate "job opportunities." For any scientist, however, research will always involve, as a minimum, teaching oneself. And the dedicated teaching of students requires a constant and often systematic search for answers to puzzles such as how best to implant an idea into a human mind. Thus, to become a biologist is to adopt a life dedicated equally to teaching and learning.

In routine activities, we get so close to our work, and to the events that seem to influence immediate decisions, that it is sometimes difficult to see the broader intellectual movements in which we participate. So there is a reason for biologists to seek out those who think in general terms—namely, the philosophers. During formal education, all seems new. The world is full of unanswered questions. Philosophers put these questions into perspective by telling us of the processes in which we are caught.

A specific problem, and the study of it, can always be placed into a historical context. The exchange of information, discoveries, and concepts that has occurred between human populations, including extinct ones, continues to influence our approach to science. For example, molecular biologists can not only tell you all kinds of reasons why they are trying to clone some gene, but can also speculate on the commercial impact of their potential accomplishment. Few will admit, however, that they are working in the alchemists' tradition. They may not realize that this (at least in part) is what they are doing, regardless of the public opinion that they are changing base elements (intestinal bacteria) into gold (*E. coli* containing a gene for human insulin). Some scholars would even say it is possible to detect a philosophical commitment to production of an elixir of life, as well as a belief that new organisms can be synthesized by changing the mix of their "elements" (DNA sequences), in the work of the gene splicers. So while molecular biology traces its history directly back to the 1920s research of Frederick Griffith (see preceding chapter), philosophers can look several hundred years back beyond Griffith and see the intellectual roots of the most modern of biological specialties. The philosophers, it seems, are not telling us what to do, but rather what we are doing.

Biology, the purposeful study of life, is a human activity; no other species studies in the same sense that we do. Biology is not a job. One may make a living doing certain kinds of biology, but biology itself is a pursuit of understanding—that is, an intellectual endeavor. Philosophers show us the nature of intellectual endeavors and suggest, indirectly, ways to best pursue them. The advice is never specific, but comes, instead, in the form of guiding principles that are themselves open for discussion: Humans are not completely aware of all the things and ways they learn, it is easy to confuse information with understanding, some people "apply" and "integrate" more easily than others, there are few, if any, right answers.

Research in the field of artificial (computer) intelligence has revealed some rather interesting characteristics of our mental abilities. Douglas Hofstadter has summarized many of these findings, and presented us with a most profound statement on the capabilities of the human mind, in his book *Gödel, Escher, Bach: An Eternal Golden Braid.* He has combined dramatic dialogs, mathematical puzzles, and recent knowledge of information-handling mechanisms, derived from attempts to duplicate human thought processes with computers, in order to demonstrate the problems associated with attempts to build an intelligent electronic device. The magnitude of those problems is a tribute to our brain, or mind (for the distinction is still not completely clear). At some time in every scientist's career, he or she must wrestle with a synthesis such as Hofstadter's. That pleasure and challenge will come to you when you are ready for it and want it. For our purposes here, however, we need to consider only Hofstadter's "essential abilities of intelligence," for they hint at the processes we use in our lifetimes of teaching and learning. In the context of a discussion of the differences between humans and machines, Hofstadter lists the following as basic intellectual abilities:

> to respond to situations very flexibly;
> to take advantage of fortuitous circumstances;
> to make sense out of ambiguous or contradictory messages;
> to recognize the relative importance of different elements of a situation;
> to find similarities between situations despite differences which may separate them;
> to draw distinctions between situations despite similarities which

may link them;
to synthesize new concepts by taking old concepts and putting them together in new ways;
to come up with ideas which are novel.

Most original study in biology requires that the learner successfully carry out these tasks. To perform them is not only to extract information and knowledge from one's environment, but also to generate one or more kinds of reality (ideas, combinations, concepts, actions). I have, of course, just defined that activity, and its products, known as research. Teaching can be thought of as the reciprocal of research; there is no reason to believe that the Hofstadter list applies less to education than to investigation.

The fate of the reality that you will generate, as well as teach, is perhaps best described by two linguists, Cavalli-Sforza and Feldman, in their book *Cultural Transmission and Evolution: A Quantitative Approach*. This work presents an analogy between the behavior of genes and the behavior of ideas in populations. Although ideas have opportunities for movement that genes do not (ideas can pass in a "reverse" direction from offspring to adult subpopulations), and the time scale for cultural transmission is different from that of genetic transmission, ideas and genes can be shown to behave in remarkably similar ways. For example, ideas that are not useful in some way tend to become extinct ("useful" here is defined in a neutral, and rather circular, manner, as being accepted and retained). The most notable example of a failed idea, and one that has been repeatedly discarded, is that acquired traits can be inherited (see Koestler, *The Case of the Midwife Toad*). Ideas that are useful multiply, evolve, and often radiate into related ones. An outstanding example of such an idea is that environmental factors can differentially influence the reproductive success of various phenotypes.

Therefore, for the biologist generating reality through performance of the Hofstadter tasks, that reality must be communicated or it will be treated in a cultural evolutionary sense as if it never existed—that is, in the same manner biological populations treat a lethal mutant. Thus if through personal study you are generating knowledge of the living world but are not communicating your discoveries, then your ideas will never have the opportunity to successfully reproduce. Fortunately, for science, cultural transmission works in interesting ways to ensure the survival of useful information. After your death, should your notebooks and records be discovered and their contents made available to the public, then your reality will become available for cultural transmission and evolution. Having previously been only the learner, you will then become the teacher. It is in this sense that historical figures as well as some current biologists are ultimately unable to separate teaching from research.

2.

What is it that is learned by biologists? Education occurs, or should occur, at several levels. At the lowest level is the acquisition of pure information, facts. Data itself is rather worthless until it is incorporated, assimilated, placed into some context. When new facts are combined with old reality a new context is produced, within which further, pure information can be interpreted. Furthermore, with development, one should eventually be able to conceptualize using data, contexts, and interpretations; to both pose questions and answer them, and to extend thought processes such as logic into problem areas not previously encountered. According to Piaget, this ability should be achieved by the age of fifteen, which also happens to be about the time young biologists become acutely aware of those naturalist values which set one forever on a path divergent from the rest of humanity.

The highest levels of learning involve what Hofstadter describes as "strange" or "tangled" loops, cycles of experiences in which one event conditions the way subsequent events are perceived, the equivalent of the mature state of Piaget's integration/assimilation stage. At age fifteen, one might buy an orchid to adorn the prom dress of a special date. Such an orchid is known to be an unusual flower, but its role is a social one and its beauty is understood only in the context of commercially available plants. Later, although a biology student may again buy an orchid for a date to a college fraternity formal, his botany courses make him pause, study the markings on the lip, and wonder what kind of insect might be attracted to them, before pinning the flower to her dress. As a doctoral student, our same young biologist may set his sights on the tropics, his goal being to compare numerous orchid species in order to formulate a theory regarding the interactions between pollinating insects and epiphytes that dwell in the forest canopy.

At this advanced stage, our student knows his observations will influence later interpretations, and knows interpretations will influence the choice of observations to make. The strange loop is the mental equivalent of being drawn into a wondrously complex jungle by the seductive lure of your own curiosity and the need to explain your experiences in broad, general terms. It is all right to follow your mind's inclinations, and, in fact, it is difficult to appreciate the full complexity of most biological systems unless you allow yourself to be influenced by what you discover about them. Without an appreciation for complexity, you are likely to never feel intellectually mature as a biologist.

Hofstadter discusses in great detail how strange loops are controlled. For example, although we may recognize the events of observation and interpretation both separately and as they interact, what we are not so often aware of is the set of rules that guides the operation of the loop itself. Think of an athletic contest. You know of two things: the general rules of play and the efforts of each team to win. But while you may understand the general rules (the controls), you are usually not aware of them in the course of play. Likewise, although your immediate awareness also includes the officials, you rarely pay much attention to them until there is a violation. Instead, most of your attention is focused on the play itself, which is almost invariably a rapid and purposeful loop of observation . . . interpretation . . . action . . . observation, each team responding to the opponent's play by modification of its own. This type of information and response interaction is no more apparent anywhere than on a basketball court.

For the biologist, the factors that control learning (rules of the game) are the ones discussed previously in the chapter entitled "The Practice of Biology." Our doctoral student is aware of what is known about orchids, their classification and geographic distribution, the ecological niches they occupy, and some of the insect species that pollinate them. He also knows what technological tools are available to help him make observations, what sociological problems (such as guerrilla warfare) might force him to modify his research plans in the tropics, and he is very conscious of his need for money. His doctoral supervisory committee has some unusual personalities; his former date, now wife, says if he spends one more semester in Guatemala she will divorce him; it is dreary March in the upper midwest; and the transmission just went out on his 1971 Maverick. He also knows that if he puts numbers, ideas, concepts, and hypotheses together properly, he will become relatively famous fairly quickly and therefore competitive for a faculty position at a fine university. These are the "rules" that will govern his intellectual growth. Now, in order to synthesize an experience to satisfy his curiosity, he has little choice but to determine a course of action in which he accomplishes his goal without violating so many rules that he is disqualified.

The highest level of learning, then, is somewhat analogous to coaching. This level involves use of the rules, or definitions of the "game," to drive the cycles of experience. For the biologist, this means being cognizant of the forces that guide the study of living organisms, then using these forces to shape one's career. Such learning skills do not come easily without studying the history of one's chosen field or of the intellectual processes that seem to operate on our heroes. In the end your goal will be to direct your learning through your own conscious understanding of the rules of intellectual endeavor. You will become your own coach. You will have learned how to learn.

After years of wrestling with complex problems such as those surrounding the pollination biology of orchids, scientists typically acquire a trait that is not a purely intellectual one—humility. At the beginning of a career one steps forward with youthful arrogance; lofty goals are set; work is of stunning importance—if not directly, then in terms of its theoretical implications—and upon the horizon stand nothing but grants, important publications, discoveries that shake the scientific world to its very foundations, and generations of admiring disciples. In the visions of some, like a distant Fujiyama, lies a Nobel Prize. All these glories may come to pass. With maturity, however, comes perspective. You will look back on your career a wiser person, aware of your place in history, and you'll make use of a lifetime of learning to describe what remains to be discovered about the world of living organisms. The humility you achieve need not negate your accomplishments, nor diminish their importance. Instead, it will finally reveal to you the size of the task you set for yourself so many years ago.

3.

What is it that is taught by biologists? Several things, many of them by example. Lessons conveyed by example may in fact completely eclipse those taught on purpose in the classroom. They may touch people you never realized were touched, affect their lives in ways you could never predict, alter your own environment in a subtle yet significant manner. Such lessons include the scientist's approach to life—that is, the reverence for understanding and truth derived from use of the scientific method. They also include values, social attitudes, the use of power, the strength of language, and the scientific community's opinion of the lay person. Notice that I have not stated precisely what values or attitudes are taught, only that some are; the specific ones taught by example depend on those of the individual scientist.

It is nearly impossible for a person to expunge implicitly taught lessons from his or her demeanor. Why? Because they are transmitted by paralanguage—by means that transcend vocabulary and grammar. Linguists tell us that in human communication, language is only the stage upon which we superimpose a paralanguage. For instance, the minute you say, "I am a biologist" you establish a context within which everything else about you is interpreted; you have set yourself apart as educated, with values not held by most. That context will validate certain of your future communications and invalidate others in the mind of a listener. By your breadth of understanding, knowledge, and erudition, you will teach your audience what a biologist is.

I have dwelt on this kind of teaching because, as I have discussed earlier, you will have a responsibility to support your own values. The extent to which you are able to carry out this responsibility depends largely on what you have taught your audience about biologists. Stated bluntly, don't expect your audience to believe your criticism of the cosmopolitan malaise of slash/burn tropical agriculture, if you have taught, by example, that biologists are narrow, arrogant, bigoted, or politically, socially, and economically naive.

You will also teach, by example, patience with, tolerance

for, and courage in the face of complexity. Of all the lessons that biologists (as opposed to other scholars) have to impart, these are among the most important. The first contact between a child and nature is characterized by wonder, a wonder that is rarely preserved through maturity. Instead, at least throughout most of the developed Western nations, nature comes to be seen by both the general public and political powers as a resource.

This progressively utilitarian view is accompanied, for some unexplained reason, by a fear of nature. I suspect we can lay the blame for this situation on para-education, the indirect instruction we receive from social interactions and the media. When a group of engineers, all working on the same hydro-electric plant, get together socially, you can bet your subscription to *Audubon* that they don't talk about the river in terms of its aquatic invertebrate communities. By the end of their evening together they will have reinforced each other's view of the river as a natural resource to be exploited. Newspapers and magazines, commercial television, and advertising of all kinds use modified formal education to convey messages. Media information is often presented in the highly authoritative package we have come to associate with classroom materials—for example, graphs, charts, and TV clips of "experiments" showing how one pain-killer outperforms another. Headline emphasis, topics of conversations with show business stars, and subjects of television shows, all convey messages of importance in addition to, and usually more powerful than, those delivered by the actual words themselves. These messages do not allow for thoughtful consideration of alternative solutions to problems, questions of meaning or interpretation, and the testing of hypotheses. Instead, they scream of simplicity and purpose, two characteristics that the natural world does not possess.

The biologist/teacher faces a general population that has experienced a lifetime of this type of education. By the time you confront the public, the majority of a random audience will feel apathetic toward the average plant and revulsed by the average animal, the average plant (large majority of known species) not being under cultivation and thus having no obvious ornamental or economic value, and the average animal (vast majority of species) being an insect. Every semester I test this contention by asking students to list, in their minds, ten different kinds of plants or animals. After a few moments of class time, I ask them to extend the list to a hundred, then to a thousand. From the front of the room I can see the smiles begin to cross their faces at the prospect of listing a thousand species. While they are still struggling to extend their lists beyond livestock, pets, and cereal grains, I end this little exercise in awareness by reminding them there are an estimated two hundred fifty thousand known species of beetles, ten thousand species of grass, a hundred thousand species of snails, clams, squids, and their relatives, and so on through several taxonomic groups. Typically, this cross section of the public considers important only those relatively few species, they think will feed them, earn money for them, or keep them company. But during the years I've been playing this little game with freshmen, the complexity of nature has not disappeared. Instead, our ability to make rational collective decisions has become dependent on an understanding of it.

The child's sense of wonder, based on naivete, is almost a form of courage; the child is undaunted by nature because he or she doesn't know how complex nature is. The true biologist, on the other hand, not only retains wonder and remains courageous, but does so after having gained knowledge of complexity. Applied to teaching, this means your audience will see you at ease with cell structure, ecosystem diagrams, metabolic pathways. Your behavior must induce audience comfort with an apparently formidable set of information. Their conclusion must be: It's all right to try to understand what I am seeing.

There is no secret technique for accomplishing this feat.

The only certain advice smacks of the negative: Arrogance, impatience, and disdain will always shut the mind of a person with whom you are trying to communicate. It helps if you admit the obvious, that your perspective is different from that of the nonbiologists. Treat them carefully, like children: Give them wonder, devise gentle ways to correct their mistakes, and do not expect them to make unsupported intellectual leaps.

Lastly, you will teach, by example, approaches to learning, to the unknown, and to life in general. Everybody does the latter implicitly; in this regard you are no different from bankers. But as a biologist you will convey the scientist's approach to the new. Throughout history, novelty has not been readily accepted by groups of humans. Although new ideas and innovations always tend to threaten some established order, history shows that in the end societies do sometimes accept them, usually incorporating them in ways that are not only consistent with the established order but that also seem, in retrospect, predictable.

Because much of what biology does generates novelty, and because this novelty is produced very rapidly, an awareness of the way society deals with novelty is important for the biologist. You cannot teach people to accept the unfamiliar immediately. Even Madison Avenue understands this. Advertising almost always ties the new to existing and pervasive attitudes and desires: Brand X is associated with beauty, Brand Y with comfort, Brand Z with success. True scientists, however, do not need such associations, for they are not likely to reject a new idea that affronts their understanding. Instead, they greet novelty with curiosity, and immediately subject it to evaluation. Plenty of new ideas do get rejected, at least temporarily, after evaluation. But in this regard you distinguish yourselves from a public that rejects the new because it doesn't like, want, or think it needs it. Thus by your reactions you will teach the scientist's approach to the unknown.

4.

How, specifically, do you become something? Clearly "becoming" involves a developmental process. But unlike the metamorphosis of an insect, this development is an aspect of intellectual growth which can begin at any chronological age. Development is also progressive, leading you irrevocably away from an original condition of naivete. How much change is enough? The answer is easy but circular: When teaching and learning about your chosen field are regular, routine parts of your life that you do naturally, without planning, or making a separate effort, then you have arrived.

Biology, unlike chemistry and probably physics, can be self-taught and practiced. The phylum Arthropoda alone guarantees this possibility, and the existence of certain plant groups increases it. There are simply so many species of insects, for example, that have not been studied in detail, that access to, and a passion for, one of them might well lead to a lifetime of study at the professional level regardless of how one makes a living. This situation is brought home to any professional who encounters what seems like a minor problem that turns out to be a significant one. I recall an instance in which a student became fascinated with not only the parasites of the darkling beetle *Eleodes suturalis*, but also with the insect itself (they are charming and survive well as adults in captivity). But her questions could not be answered unless she could raise *E. suturalis* from eggs in the laboratory. Her captive adults laid hundreds of eggs, and a few hatched, but the larvae would not live no matter what she did for them. She inquired of a leading professional entomologist about methods of culturing the species. His response was encouraging because it indicated her failure was not unique and that the species was virtually impossible to raise in the lab, at least with present knowledge. I remember thinking: Here is a great problem for some amateur with patience, insight, and a love for *Eleodes suturalis*. A description of the culture methods alone would not only merit a fine publication, but also make this species available for a variety of developmental, ecological, and behavioral studies. This same story could be told, simply substituting the appropriate scientific names, about literally thousands of species of plants and animals.

While the areas of biology available to the self-taught scientist may be somewhat restricted (you'll probably not become an amateur cell biologist), the level of practice is less so. Some species which have been largely ignored can be studied in great depth, and the resulting discoveries will bring as much satisfaction to the amateur as to the professional. At a minimum, however, one must have access to two resources: the biological material and an adequate library. These resources are provided in part to enrolled students as the function of most universities, along with the faculty necessary to award degrees certifying that the students have successfully studied their coursework. Certification in the College of Experience is not so easy to obtain, but in the long run it may be just as valid.

In any field the education of an amateur (by "amateur" here I mean one who is not employed as a biologist) differs from that of the professional in a number of respects. First, and most important, the amateur's learning experience is not coupled to the academic or industrial reward system (grades, scholarships, tenure, money). The amateur is often free of the pressure associated with these systems, but has restricted day-to-day access to other scholars, including energetic young professional biologists. These young biologists have a way of evaluating the leading thinkers of their day, creating an intense intellectual swirl which usually doesn't include the amateur. There are people who think that in this respect the amateur is better off than the formal student, since original thinking comes surprisingly often from those who are free of pressures and influence.

Second, in any field of science, intellectual stigma is sometimes attached to amateur status by the professionals. While this attitude was present even when biology was largely an amateur business, it became more pronounced after World War II because of the rapid introduction of chemistry- and physics-based technology into biological research. There are exceptions, of course, but in general, practicing professional scientists simply look down on amateurs because the latter so rarely make significant discoveries in research areas that are currently emphasized. The sociobiologists might speculate that such a condescending attitude is a form of either dominance or group defensive behavior. Nowadays, with the sciences so dependent on government support, it would seem to be a wise political move for the professionals to try to make an amateur biologist out of everyone.

Today's ecological awareness has made major inroads into professional snobbery. Organizations such as Earthwatch recruit amateur volunteers to serve as research assistants, citizen environmentalists learn to speak the ecologists' language to pursue their goals, and recreation and organized travel with a heavy biological emphasis is proving a profitable business operation. In other areas, poor economic conditions have strengthened the importance of amateur ornithologists and botanists in the eyes of the professionals. There is simply not enough public money available to gather the original observations needed, for example, on annual bird migrations and population estimates. But don't expect to develop into an accepted amateur molecular biologist unless you have the chemistry, physics, and math background to generate original theoretical contributions with pencil and paper. There are people out there, incidentally, who are capable of doing just that.

The amateur, unlike the professional, is also totally free to explore personal interests without regard to formal publication. This difference is a rather critical one, mainly because publishable science has some fairly stringent prerequisites regarding data analysis, numbers of observations, experimental design, and the like. These prerequisites are not so restrictive that they would prevent an amateur from being a true biologist. They can, however, be restrictive enough so as to steer an amateur into those kinds of biology from which one can reap all the rewards (pursuit of personal interests) with none of the hassles (grant applications, publication quotas in "good" journals).

What do amateurs "do" with their education? This question bears asking because of widespread parental pressure for young people to get an education they can do something with. Somehow employment has come to validate the pursuit of a college degree. The amateur does not make a living as a biologist, but he or she can help increase public awareness of ecological issues, influence politicians to make reasoned decisions, offer educational experiences the public has chosen not to pay for, and provide the example of a person vitally interested in the world of life. Such "products" can be of more real value to society than whatever is provided by one who is making a living, and nothing more, by doing biology.

Furthermore, the exploration of nature out of purely amateur interest has produced some astounding contributions to biology. Charles Darwin was, in essence, an amateur; so was Gregor Mendel. Walter Rothschild, of the English branch of that monumental family, started a natural history museum at the age of seven. By the time he died his Tring Museum collections rivaled, and in some areas exceeded, those of the British Museum of Natural History. Single-handedly, through his pursuit of personal interests, he left us a legacy that will take generations of scientists to explore thoroughly.

The ultimate question to ask, if you consider yourself an amateur, is: Do I feel I am developing intellectually, departing progressively from naivete in a manner that pushes me inexorably into a lifetime of teaching and learning about organisms?

If the answer is yes, you are probably on the right track. While you may not be as well endowed as Walter Rothschild, you can nevertheless be a serious student (in the general sense) who, except for the financial remuneration, is a biologist. And everybody knows you don't do biology for money anyway.

5.

There is a common right of passage performed by most of those who would be biologists—namely, that of graduate school. Graduate students in biology can frequently pick one another out in crowds, empathize, degenerate into esoteric conversation, and adjourn to the nearest bar without the formality of introduction. Poverty alone doesn't seem a sufficient explanation for such mycetozoan behavior; there must be some other shared experiences, perhaps the struggle with data, techniques, publication, and mentors. Graduate students are in the midst of the most intense learning experiences of their lives. When they meet in their common zeal, they exchange a vast amount of information that helps each take another step toward success. At no time, and under no conditions, are teaching and learning so merged as when grad students gather.

Then why is not everyone in graduate school? The most obvious explanation is that employment opportunities for specialists tend to vary with the tides. At the end of a decade of formal education one faces a world that asks, "And now that you're thirty years old, what have you been doing all this time to qualify yourself for a useful role in society?" There is an inescapable feeling that the moment of truth has arrived when all you can answer is, "I've just spent five years gathering the information to write a six-page paper entitled 'Epidemiology of *Plasmodium hexamerium* Huff 1935 in Meadowlarks and Starlings of the Cheyenne Bottoms Waterfowl Management Area, Barton County, Kansas.' "

Obviously, serious students can often use some psychological support. Somebody outside academia needs to agree that no matter what the immediate future holds, somehow, maybe by unexplained processes, graduate work will in fact have been worth the effort. In addition, a grad student must develop an outer toughness which dismisses the feelings occasionally expressed by others that he or she has been avoiding employment by writing a thesis. The comments that follow are opinions and ideas, not hard-and-fast rules or facts, that may provide some support, or at least sympathy. They are based on observations of a great many graduate students, so at least should allay any anxieties that one's own fate is somehow depressingly unique.

The biggest problem a graduate student faces is The Problem, the one which provides life focus and identity—namely, the thesis or dissertation project. There are both idealistic and practical considerations involved in the handling of The Problem. Idealism is one of the most foolish luxuries in which academic scientists indulge themselves; for a graduate student, it can be deadly. Idealism can take many forms at this stage in one's education—for example, commitment to a research project which is far too ambitious, determination that one's thesis will be the finest ever written, a belief that the dissertation is the end, rather than the beginning, of a research career.

I am reminded at this point of some decidedly simple yet brutally honest and practical advice I received at the start of my doctoral work. I was on a tour through the National Institutes of Health in the tow of my mentor, Dr. J. Teague Self. I had at that point been Teague Self's graduate student for about two weeks. Walking down a hall, he cocked an ear at a blustery Irish voice emanating from an office, poked his head inside, then turned back to me and said:

"How'd you like to meet Coatney?"

Of course I said yes, and was ushered into the office, gestured into a chair, then left in the company of G. Robert

Coatney, A. M. Fallis, and J. F. A. Sprent, men whose names I knew from reading of their research in the literature, and whose faces I felt, because of my respect for their work, should be carved into some mountain in the Black Hills. I can think of no time since in which I've seen any other student so suddenly flung alone into a room full of his heroes. After a few moments' discussion of graduate school and bird malaria, during which all three men seemed to be transported back to some days they remembered fondly, Coatney issued the advice which put all idealism into perspective, the hindsight of a long and successful career, and which spoke of a hazard never imagined by the young.

"Always," said G. Robert Coatney, "be finishing something."

Those words are a haunting indictment of a certain approach to life in which pursuit of an unreachable goal, no matter how important it may seem, consumes a person's creative powers yet never allows one to take the next formal step toward professional maturity. Although it may sound like blasphemy in some circles, there is nothing wrong with choosing a thesis project that can actually be done within a reasonable length of time. If you are anything at all like the rest of us, your dissertation will not win a Nobel Prize, save humanity from itself, get you tenure a dozen years down the road, make you rich or famous, or guarantee you a job. It will, instead, be an intense research and learning experience that teaches you how to do the business of biology. Coatney's advice rings sharp: Choose something you can finish. If your famous mentor persists in throwing destructive idealism in your way—for example, continually adding questions for you to answer so that after a few years you seem no closer to the end than when you started—change mentors.

I will define a valid graduate thesis problem, then, as one that can be done with the time, money, equipment, and expertise resources that are at your disposal. This sort of validity is of a practical nature and in real life overrides biological validity. At the graduate student level, many reasons can be put forward to justify a struggle with the logistics of research. Such a struggle, the more strenuous the better, helps ensure that you will not enter the market helpless as a thinker. The most devastating situation a young scientist can face is to suddenly find himself or herself employed, expected to produce, yet wondering where to start converting ideas into discoveries. Suddenly all your mentor's grant-supported technicians, dishwashers, animal caretakers, greenhouse gardeners, stockroom personnel, and secretaries begin to come into focus in their true roles. Suddenly the famous mentor's name, if it graces your signed thesis, seems sadly inadequate.

On the other hand, practicality can lead a person into triviality. In the extreme case, a mentor brings a student into the lab, assigns a Problem that has its pre-ordained place in the Grand Research Scheme, makes the student depend on available equipment and established techniques, and virtually guarantees publication. Beware of such comfortable situations. They will not provide a graduate education, only certification. The one ingredient they lack is idealism. The Problem, then, should be one whose solution requires a mix of practical challenge and idealism that works with the student interests, personality, and specific mental equipment to produce a certified and educated biologist. While the mix is brewing, the student must make mistakes, suffer the consequences, correct them through his or her own devices, reap the pride that results from having done so, and finally receive the credit. Mentors and schools that do not allow this flexibility and excitement should be rejected by the student.

The finest graduate experience is filled with thrills and spills. The thrills are a growing sense of understanding, intellectual maturity and strength, personal discovery, and the confidence that one is able to solve a Problem. The spills are equipment and utility failures; odd weather years; erroneously mixed reagents; bad advice (from yourself and others); various lab mistakes and procedures that introduce variation into data, personal problems (wife, husband, children, affairs, alcoholism and drug abuse, poverty, old cars); occasional faculty jerks; university rules, regulations, and policies; and academic politics. Obviously, the spills, no matter how devastating they seem at the time, aren't in quite the same category as the thrills.

6.

Does formal education actually prepare one for a job? Generally not. As I pointed out earlier, formal education certifies a person, who by virtue of that certification may find employment, thus gaining the opportunity to see if he or she can actually do a job. In the case of graduate education, dissertations alone rarely qualify one for much of anything. Along with coursework, they are supposed to guarantee a potential employer that a student has acquired at least the basic equipment a person brings to any task: information and technical skills. A Darwinian conclusion follows: That person who has acquired the most broadly applicable information and technical skills should have the least trouble converting formal education into employment. For example, all other things being equal, the brand new molecular biologist who can, and is willing to, teach nursing physiology or elementary microbiology will probably find a home in academia more quickly than the one who disdains these challenges.

Extensive education in the sciences prepares one for a job that should have, as an integral component, the requirement that a person continue to become educated. Thus, when one finds employment, teaching, learning, and practice ideally become merged. The routine image of professional employment includes the elements of commitment, tangible rewards proportional to performance, and advancement. The first lesson a would-be biologist learns after formal education is that rewards are not always presented in dollars. They come in the form of successful research projects, good ideas, exciting experiments, personal insights, and colleagues who recognize your accomplishments. The second lesson is that biology, like virtually all professions, actually restricts one's access to the world of life. Maturing in a career demands continuing achievement, which can be gained only through focused attention. For a biologist, the organisms encountered, the manner of dealing with them, the questions that grow from experience may become progressively narrowed in the course of one's work. Some welcome this restriction, others find it depressing.

The third lesson is one of the most surprising and cruel: With your first advancement you are suddenly asked to perform tasks for which you were not specifically trained. This principle puts the practice of "going to school in order to get a job as a (something)" into a rather jaundiced perspective. For example, only rarely does dissertation research prepare one to be an effective committee chairperson, supervise others, manage grant accounts, design curricula, or cooperate with professionals in the business world. The unhappy truth of this statement is behind the lament of many senior scientists I know who, sitting in the lobby of a hotel at some national meetings, observe that their organizations have hired highly specialized young researchers for a decade and suddenly, "Dr. X has retired, Dr. Y has taken a job elsewhere, and there's no one left who can run the department or teach freshman biology."

For anyone who anticipates an education that will lead to success as society normally defines the term, Mortimer Adler's *The Paideia Proposal* is required reading. Adler addresses education by defining those human skills it should develop: communication, problem solving, and poise when confronted with new information. These skills are manifest in those mental activities we know as reading, writing, and mathematics. Again, it seems, we appear to have reached a trivial conclusion, for obviously, a person must be able to read, write, and figure in order to function. But in the course of his analysis of this apparent triviality, Adler makes two very astute observations. First, he points out rather forcefully the impressive extent to which these skills are not only cross-professional—that is, common to a variety of seemingly unrelated jobs—but are also critical to any society that depends on relatively full development of individual potential. Second, he makes a strong case against vocational training—that is, going to school to prepare oneself for a specific job. Vocational training, Adler argues, inhibits development of communication and problem-solving skills because it does not assume professional growth. In this curious way, vocational training limits, almost by intent, a person's ability to adapt, advance, conduct the biologist's required life of teaching and learning. The truth of Adler's argument is borne out by current trends in business. Information processing needs have stimulated changes leading to increasing sophistication in equipment and communication systems. As a result, industry today is heavily involved in continuing education of employees at all levels.

How does Adler's analysis apply to biology? It suggests that the goal of formal education should be to acquire cross-professional skills rather than specific training. For example, if this conclusion is correct, then a biologist should be able to identify, using appropriate literature, organisms in taxonomic groups with which he or she is unfamiliar. After all, the approach to taxonomic study, the manner in which literature is used, and the habit of evaluating characteristics should be broadly applicable, while detailed knowledge of structure and vocabulary can be acquired as needed. Similar transfer of investigative and communication skills—that is, the biologist's analogs of reading, writing, and figuring—should be possible in most areas of biology such as physiology, ecology, and behavior.

Although, of course, no Ph.D. candidate is in vocational

training, the vocational approach to education is one that a graduate student must continually fight, mainly because we are so often inclined to equate quality with specialization. Doctoral supervisory committees may think they're making sophisticated and well-informed decisions by approving programs that are intended to accomplish certain defined goals: providing research tools "appropriate to The Problem," converting a person into a community ecologist, forcing a reading knowledge of a foreign language in which is written most of, say, the "literature on the tapeworms of reptiles." Despite the specializing tendency, however, Adler's concept of education generally holds true at all levels of human affairs. The less specialized person has the most potential for fully exploring his or her interests, limits, and possible roles in society. With that perspective, the specific advice to offer a student then becomes: Accede to the minimum vocational training required to certify yourself, then invest the rest of your efforts in breadth. In the long run, education intended to produce a molecular geneticist, a systems ecologist, or a cladist is inferior, both for the individual and for society, than that which is intended to produce a broadly educated person who has also written a dissertation.

7.

For the young scientist, few pieces of literature are more illuminating than LeShan and Margenau's *Einstein's Space and Van Gogh's Sky* (see section 8 of the previous chapter). In this book the authors distinguish among the different domains of intellectual endeavor, including those of art, social science, ethics, science, and parapsychology, and describe the methods for establishing reality in each. For example, feelings may be used as a form of reality by the arts, but are not accepted in the sciences. In addition, the laws which operate in one domain may be totally unrelated to those that operate in another. For example, an artist may produce an abstract expressionist painting by somehow communicating with an unidentified inner source of images, but a scientist had better be able to measure objects or events in order to test a hypothesis. LeShan and Margenau point out that, depending on the domain, emotions, experience, impressions, concepts, equations, measurements, intuition, and ideas can all be considered real. Normal science as applied to biology, with its dependence on physical measurements of the tangible realm, occupies but a small fraction of this spectrum of reality.

Inherently and by definition, a broad education demands a person think by the laws of reality which operate in different domains. A student should at least be exposed to the rules and practices of social scientists and artists, as well as those of normal science. Why is such exposure critical? Because the actual human experience of teaching and learning crosses into multiple domains. Scientists themselves step out of the narrow realm of measurable phenomena whenever they begin to theorize, speculate, and practice the art of educated insight, activities they do regularly. Science continually borrows methods, perhaps unconsciously, from the reality domains of other disciplines. For example, although published papers carry the aura of neutrality, they are in fact heavily infused with personality through choice of words, word sequences, graphic designs, and the very journals in which they occur. The papers' authors operate, at least while composing, with one hand in the world of the novelist. Similarly, lectures, seminars, and symposia are forms of theatrical productions, and so result in the transfer of information according to laws which operate in the domain of drama. Yet we rarely question our impression that the education we receive from all these sources is indeed in the realm of science.

The most effective pedagogy involves the use of several

learning modes—for example, lecture, demonstration, experimentation, and role playing—to reinforce a given idea. Such methods rely on the use of operating rules from several domains to convey a concept that is intended to apply in a single realm. Imagine that we enter a darkened room to teach landlocked midwestern freshmen, 4-H still coursing in their blood, something of the architecture of sponges. We choose the most beautiful and dramatic pictures of spicules, relate the secretion of species-specific skeletal elements to the aging and sexing of hominid fossils, discuss evolution of body type in parallel with the embryology of a pig's lung to illustrate folding, and in the end generalize, using the rhetorical devices of an actor, by asking how a principle of development derived from sponges could be so applicable to our lives, how so often the true nature of something is obscured until we study its ontogeny. In the silence, we spread the leuconoid principles across individual lives, across history: How did we get into this condition? We got here through a progressive series of changes. No one in the class has seen, touched, measured, or chemically analyzed a sponge, yet all feel they've learned. Given the fact that they can't walk directly from the lecture hall to the ocean, they have. The subject is biology.

That the subject is biology is critical to our discussion. Biology is the study of life; that can be said many times without losing its force of reminder. We do not even know all the species on Earth, much less understand how they came to make their peace with the planet—information we ourselves could put to good use. In the tropical forests we are destroying them faster than we are discovering them. Day by day, published research only increases the complexity we see in these organisms' relationships to their environments, to one another, and to us. By virtue of the subject matter, one who would study life can only learn to learn, to teach oneself the many paths of understanding and the tools to clear the way. Surprisingly often the paths lie in domains other than science, and the tools are those of artists, historians, actors, and psychologists. The fundamental gadget of science, of course, is the hypothesis, but it alone will make one as much of a biologist as the piano makes the composer, the turpentine the artist, the office blackboard the mathematician. For the individual, the task at hand is clearly the merger of research and teaching in their broadest sense as ongoing and complementary properties of a chosen intellectual way.

4.

MAKING

A LIVING

> . . . we are meant for work and government,
> for austerity: and these shall take
> priority over love, dreams, the spirit,
> the senses and the other second-class
> trivia that are found among the idle and
> mindless hours of the day. . . . Damn
> them, they are wrong.
>
> —THOMAS PYNCHON

1.

Making a living means, for the average person, earning money. Most scientists, including biologists, face a lifelong financial struggle, since fiscal survival means not only grocery and apartment money but biology money as well. Although physicians, dentists, and veterinarians do much better financially, they are not, by definition, biologists, regardless of what is on their undergraduate transcripts. Ironically, money often becomes more of a problem if a biologist's career progresses successfully, if for no other reason than that continued productivity requires a continuous search for funding, an activity that can come to occupy a major portion of a scientist's time and thoughts. For the vast majority of scientists, time eventually leads to a compromise between the pursuit of intellectual interests and the quest for money. One always inhibits the other.

At first, it may seem trivial or obvious to mention ways and places to get money: marry it; inherit it; find employment in academia, industry, or government; play the stock market; be something you don't want to be. It is not, however, a trivial experience for a young person to ask himself or herself seriously where the funds will come from to support a life's work.

Charles Darwin married money, in the form of the granddaughter of Josiah Wedgwood, the original manufacturer of Wedgwood china. As a result, Darwin was largely isolated in his day-to-day work from academic politics and infighting, the rules and practices of academia, and the necessity of working for someone else in order to support his own explorations. Anyone who has experienced the effects of fractured attention on focused scholarship can understand instantly the contribution Darwin's wife made to biology.

Since marriage came fairly late in Darwin's career, after his voyage on the *Beagle*, he brought to the relationship the bulk of his field work complete. This was not the case with John James Audubon, whose wife would have welcomed a fortune equal to that of the Wedgwood family. Audubon spent so much time, and traveled so far and wide, trying to peddle subscriptions to his folios that it is a wonder he was able to produce as many paintings as he did. His biography reads more like that of a costermonger than a biologist, yet today his name is not only associated with a body of work that provides a powerful view of the natural world, but it also graces the cover of the publication that most strongly promotes the naturalist tradition.

Sir Walter Rothschild inherited money. His niece's recounting of his almost surreal life (*Dear Lord Rothschild*) is highly recommended reading for any young biologist. Sir Walter's unique combination of resources, intellectual and financial, enabled him to make a contribution to biology that shares one characteristic with Darwin's: It has kept, and will keep, generations of biologists occupied. While Darwin's legacy was conceptual, Rothschild's was material—namely, his collections. By the age of seven, he had decided to have a museum. By the time he was thirty, family funds had built not only Walter's Tring Museum but also homes for its curators. In the years that followed, the Rothschild fortune financed worldwide collecting expeditions; the salaries of curators and their assistants; the maintenance of live kiwis, cassowaries, zebras, and giant tortoises; travels; taxonomic research; the publication of forty volumes of *Novitates Zoologische;* and two black-mailing mistresses. Out of a sense of family responsibility, Walter had tried very hard to do what many biologists today must do for money by being something he didn't want to be —in his case, a financier. After eighteen years behind a desk, he admitted defeat and went back to his butterflies.

Inherited wealth allowed Walter Rothschild an option many biologists, withering away in university administrative positions, do not have: an exclusive devotion to the nonhuman world of organisms. Advancement through the ranks of any organization is certain to take a true biologist further and further from the organisms. Standing in a World Health Organization building in Geneva, I once heard some of the saddest words ever uttered by a scientist. His name was one I'd known from the literature long before we met as part of a Scientific Working Group advising the WHO Special Programme in Tropical Diseases. I had been shocked, at first, at the disparity between his age and reputation. I'd expected a much older, perhaps bearded eminence who might, if treated with proper respect, speak to me about parasites. Instead, I found a young man who could have passed for an untenured assistant professor at any midwestern university. But as our service on the committee extended over several years, he revealed the intellectual strength, objectivity, perception, and analytical power that had helped to spread his ideas around the world so quickly. At my last meeting with the group, he announced he was giving up research, the last thing any of us expected to hear. Privately I wondered if he was hiding knowledge of the earliest stages of a terminal illness. Then he explained. He was being promoted to Full Professor, an advancement he could not refuse. With the rank, however, came teaching and administrative duties that would consume all his energies.

"In our country," he said, "it is the law."

I realized then that it was the same law that operates in all countries: Promotion takes you further and further from your roots, from the crawly world that first captured your life's attention, and plunges you into the world of human affairs. We are speaking here of an especially cruel illustration of being something other than what you want to be in order to make a living. It happens every time a true biologist becomes a department chair, a dean, a chancellor, a director of research. Biologists handle organisms; directors handle money and other people's careers. A biologist who does his or her work rapidly and well stands an excellent chance of facing the choice to stay with a life's work or "advance" into another position. Anyone who thinks this will be an easy decision is a fool.

Since I have no knowledge of the stock market, I'll try to pass along some ideas about the remaining three major areas in which most biologists find support: academia, industry, and government. Each has its advantages and drawbacks. As in all fields of human endeavor, these properties can be expanded, overridden, canceled, exploited, or fatal, depending on an individual's creative powers and abilities to adapt. What is a major drawback for one person may be a screaming opportunity for another. That fact alone should make one wary of specific advice.

2.

No life, except that of the independently wealthy, can match the freedom enjoyed by an American college or university faculty member. Your own enthusiasm can generate excitement in hundreds of students who in turn will stimulate you.

They can be fascinated by the most obscure beetle or drive you to seek understanding of some process you've tried to ignore. They will return to thank you for things you made them think about. You can choose to mark lizards, sequence DNA, or map the distribution of pine trees and someone in academia will understand why you are doing it, appreciate your work, and want to discuss your discoveries. Tenured academics are a truly privileged class. An academic position is a goal worthy of any young biologist's ambitions.

Faculty jobs today virtually require an advanced degree, usually a Ph.D., and often post-doctoral research experience as well. Depending on your area of specific interest, you may be competing with hundreds of others, or fewer than a dozen, for a faculty position at a college or university. The starting salary is not likely to be high. Appointments are usually for nine months, or the academic year. Fringe benefit packages vary from poor to exceptional. (The poorest ones include features such as restricted or nonexistent travel funds and group health insurance negotiated annually and awarded to the lowest bidder; the exceptional ones may include tuition waivers for dependents.) Young biologists, on the whole, are notoriously naive when it comes to salaries and benefits—after all, education in these matters is left to the College of Experience. In general, for those who are materially oriented, academia will seem like an ideal place in which to get ripped off. As a document that proves the intangibility of academic life's rewards, this book falls far short of any university's printed budget.

Institutions vary so much that it is difficult to generalize about their attributes as working milieux. For one person, a large, prestigious university represents the ultimate intellectual environment; for another, such places offer mainly bureaucratic hassle. It is a rare academic biologist who does not teach, yet teaching obligations can range from devastating at some smaller schools to schedules, at a few larger state universities, that you are ashamed to discuss with taxpayers. If you hate to teach and are bad at it, any teaching is too much, and you don't belong in academia. If you fall into this category, most small schools won't allow you to remain on the faculty for very long. Larger universities, in contrast, put a considerable demand for research on young faculty, and tend to promote on the basis of research performance, hiding or rationalizing poor teaching if it threatens the school's research reputation.

In approaching an academic life, it might be wise to consider the following: It is almost impossible to stay intellectually alive without doing some kind of regular creative or problem-solving activity. Perhaps I've just said that for the benefit of administrators and legislators who may have picked up this book, thinking it might suggest ways scholarly work in philosophy or musicology can benefit agriculture and promote tourism. Academia will exert little pressure relative to the kinds of research, or the specific projects, you pursue. On the other hand, pressure to actually carry out these projects, publish the results, and find the money to do more can be intense, especially at prestigious institutions or ones trying to become so. The more of this kind of pressure you have to deal with, the greater your need for a deep and basic love of your chosen material, communication skills, self-confidence, and a sense of the history of science. These qualities will operate to keep you working when you're frustrated and exhausted, to obtain money for studies with no immediate economic payoff, and to assure you that people in worse situations have ended up making major contributions.

The only major negative element in academia, and one that cannot be overlooked, is politics. Daniel Greenberg in his book *The Politics of Pure Science* cites a line from Jerome Ravetz' paper on the Mohole project (a plan to drill into the Earth's interior): ". . . the norms of social behavior appropriate to politics, business enterprise, or even speculative technology are not those of science. When they are imported into science, they constitute corruption." No clearer statement could be made to characterize the behavior of university faculty members fighting over reputation. Common and, incidentally, accepted political behavior includes vote trading, favoritism, subordination of performance to organizational loyalty, purposeful bias in representation, secrecy, decisiveness in the face of complex issues whether that decisiveness is born of understanding or not, blaming opponents for problems that may be long-standing and of a deep and fundamental nature, and the intentional failure to solve serious common problems because of potential voter dissatisfaction. Common and accepted business practices include efforts to sway opinion through advertising or to influence financial decisions through manipulation of image regardless of the true nature of things, secrecy, lobbying, talent raiding, litigation or the threat of it, use of a bewildering array of laws to present through accounting a particular version of the truth, the hiring of skilled accountants to do the same, and the subordination of content or analysis to monetary considerations.

These norms of business and politics are not the expected behaviors of a biologist searching for mechanisms by which gene expression is regulated. One would also like to hope they are not the activities of a faculty deciding whether to recommend our biologist for promotion to associate professor with tenure. There are few who have been in academia a decade, however, who have not witnessed many if not all of the above practices. In previous sections of this book I have tried to illustrate various points by using anecdotes, but to do that here might well stir embers that are best left to cool. Let us just say that when, by virtue of an official assignment, one has power over others, and when people try to make group decisions, individuals who are involved at the time have a way of later looking back on events and wondering if the group really took the correct course of action. The ultimate general reference on this phenomenon is John Gall's *Systemantics*.

Reputation is the currency in all of science, as well as in

all fields characterized primarily by products of the mind. This fact may be emphasized out of proportion in academic settings. Academia is almost by definition financially poor compared to industry. Business executives are typically dismayed by academic squabbles and power struggles; there seems to be absolutely nothing at stake. What such outsiders don't understand is the emotion of people fighting over ideas. Experience demonstrates repeatedly that such conflicts are among the most bitter and least flattering to the species. One only needs to read a history of the Crusades in order to become convinced that this contention is valid. In academia, the fighting over intangibles often elicits very lowly behavior, quite unexpected from a bunch of the most highly educated people in the world, especially ones who claim as their life's work an objective search for truth.

You may be wondering by now why academia, poor and vulnerable to internal stress, would appeal to anyone. For the most obvious answer, you need only return to its potential for intellectual freedom and time to pursue your ideas. The human is first and foremost a product of the mind. Nowhere outside academia does one find such a wealth of opportunity to participate so fully in the collective mental development of our species. The university offers you an open chance to test your single mind against those of all humanity. The progression from curiosity, to a decision to study, to research, to publication, to the development of an extensive line of thought about the natural world, can be experienced nowhere else as regularly, and with minimal interference from distracting influences, as in academia.

In addition to the intellectual freedom, only at a university can one participate in the highest level of teaching, that of directing the work of graduate students. With success at this level comes the opportunity to say, "I have helped someone to become a serious thinker, I have opened doors that could so easily have remained shut, I was brought a human mind and through my efforts saw it mature into that of a true biologist." One wonders, finally, if academia remains in such precarious straits, compared to business and industry, because everyone realizes the things I've just listed are things that cannot be bought.

3.

From now until the end of the century, industry may be the "hottest," and in many ways the most exciting and challenging, environment for a biologist. A serious look at industrial career opportunities in biology today requires first an examination of the history and sociology of science, beginning with some revolutionary events of the last decade or so.

This is the period in which large numbers of people first began to realize that exponential growth of the human population was actually challenging the fixed resources of Earth. Science realistically predicts the exhaustion of our fossil fuel supplies and with it our ability to make fertilizer. Massive malnutrition, heavy expenditures on arms, an ever-growing population of illiterates, and global climatic changes resulting from human activities are added to the mix of conditions in which you are deciding to become a biologist. The magnitude of the problems dictates intellectual ventures that as recently as the 1960s would have been considered scientific heresy or at best fiction. Industry now sees in ecological catastrophe an opportunity for monumental fortune, provided a company can develop products to even temporarily offset Armageddon by devising inexpensive ways to produce fertilizer and vaccines, repel agricultural insect pests, and exploit the numerous plants whose natural products have medical uses. The field that has produced this vision is molecular biology.

The dramatic speed with which this field has grown is illustrated by the following: In 1974 the city of Cambridge,

Massachusetts, fearing the accidental production of a bacterial Frankenstein, passed a city ordinance against recombinant DNA research. Two years after that historic expression of distrust, Robert Swanson, a venture capitalist, and Herbert Boyer, a Stanford scientist, formed Genentech, a company whose potential was dependent on its ability to manufacture "natural" products using genes transplanted into bacteria. The company went public in 1980 with an offering of thirty-five dollars per share. That same year, Walter Gilbert of Harvard won a Nobel Prize for his research on DNA. His company, Biogen, went public in 1983 at twenty-three dollars per share. By 1984, Gilbert and his wife owned twelve million dollars' worth of shares and he drew an annual salary of $274,200.

These are rather heady circumstances for biologists, and one has little choice but to ask how and why they came about. The situations actually rest on three different aspects of molecular biology that are familiar to all who read the newspapers: (1) the development of techniques for isolation and insertion of genes from a wide variety of organisms into bacteria; (2) the relative ease with which some micro-organisms can carry out certain syntheses and degradations—that is, in comparison to humans working either through their cells or in their laboratories; and (3) the high commercial and social value that has come to be attached to biological products such as insulin, growth hormones, and antibodies or antigens for vaccination and immunodiagnosis. Our requirements, and in some cases desire, for biological products emphasize the naturalist's perception of how tied we are to the planet upon which we evolved. That we ultimately live under the same circumstances as other organisms, prone to disease or disability and in constant need of food, is not only a fact that underlies the commercial success of molecular biology but also one that humans have always tried to hide from, ignore, or escape. Molecular biology has given us renewed hope that we might just carry it off. With the hope comes industrial money.

The period from the mid-1970s to the mid-1980s saw sophisticated research procedures in biology become almost commonplace technological tools. This sort of transformation occurs often in science, but rarely has it happened so quickly, so visibly, and in such an economically optimistic climate. The result is an almost overnight demand for people, from trained (and trainable) technicians to research scientists.

On the surface, industry could hardly look better as a place for a modern biologist whose credentials include specialized training in a combination of microbiology, biochemistry, and genetics. Industry generally pays well, certainly compared to academia; working conditions for college graduates are usually excellent; benefits are reasonable (after all, industry has been seriously confronting the benefits issue for decades); and employees at the professional level often have a sense of mission and togetherness missing in a college faculty, perhaps because companies may pursue a rather specific set of goals and people are not usually hired unless they fit into research, development, and marketing strategies. But industry also does a number of things academia does not generally do: move to exploit new discoveries with economic potential, compete with similar institutions on a short-term basis, and adjust its work force to accomplish these ends.

It is the latter tendency which will most influence the career of an industry biologist. Work forces can be adjusted through changes of personnel or changes within personnel. As a result you may find yourself working on problems that employ your technical skills and insights, but that are part of a research and development plan devised by someone else. A personality that demands intellectual independence above all else may have some difficulty with this arrangement. Again, it seems, it is worth taking a close look at the kind of person you really are, assessing whether you will fit into the environments in which you are seeking your future, and adjusting your dreams accordingly.

In a general sense, then, industry offers less freedom to choose your study subjects, but better pay, benefits, and facilities than academia. Companies also differ significantly among themselves, truly large ones having the capital to support long-term research and development, but smaller ones sometimes being unable to invest heavily in these activities. As a result, a biologist may end up concentrating his or her efforts on the final stages of product development and packaging in a firm that cannot afford research. Industry is often unfairly presented as greedy, self-serving, and interested only in short-term gains. Businesspeople, especially those in established leadership positions, are not stupid, but neither are they in business to lose money. If a company is to survive, the long-term investment must eventually pay off. For this reason, ideas usually must survive scrutiny by research and development administrators before a company will make a long-term commitment of resources to them. Once again, however, the history of science shows us that projects with developmental potential are usually those upon which academia has already done the real groundwork and gambling.

Development, the conversion of a discovery into a marketable product, is one activity that requires biological expertise but has no counterpart in academia. For example, the story of a single chemical compound, from discovery of its biological activity to its dispensation by a pharmacist, gives an interesting picture of the range of employment opportunities in industry. Primary screening of the substance may involve testing its efficacy against organisms themselves, or against infections in lab animals or cell cultures, or, in the case of hormones, testing their performance in bioassays. Secondary screening involves tests for the action of the compound in animals physiologically closer to humans than primary screening models such as mice. Determination of the compound's biological activity only begins its journey to the shelf. Toxicity testing, formulations for dose regulation, development of treatment regimens, formulations to prolong shelf life, discovery of synergistic effects, description of physiological side effects, and development of specific targeting methods all require significant amounts of biological and medical expertise. Although actual field testing may involve physicians if the product is for human use, a drug that reaches the hands of an M.D. has seen mostly biologists and chemists for the decade it takes to bring it to the field-testing stage.

Of course industry includes more than just the recombinant DNA business. Hopes for the near future, for example, include the use of artificial DNA in computers for information storage and retrieval, the basic function for which the molecule evolved in the first place. At the other extreme of esotericism one finds the cosmetics industry. Every new flavor of lipstick must be developed and tested in a manner similar to that of pharmaceuticals. The prosthesis business likewise consumes large amounts of biological expertise, although obviously much of it is medically oriented and based on biophysics. Prosthetics today goes well beyond the wooden leg. While Barney Clark's artificial heart was more a demonstration of our cultural and economic status, and our concern for cardiac problems (as opposed to starvation or parasitic infections), than a scientific achievement, it also brought into focus the level at which we are trying to provide artificial parts and functions. Although the heart business became high drama in this country with the elevated standard of living after World War II, less visible but perhaps more ambitious and sophisticated prosthesis development now centers on artificial hearing and eyesight and computer-controlled movements for the paralyzed.

Working in parallel with the prosthesis industry are the organ transplant efforts of modern medicine. The major problem with transplants, of course, is that they elicit immune reactions. The changing sociology of transplantation, and our growing acceptance of it as a solution to medical problems, have produced not only donor cards, something your great grandparents could never have conceived of, but also a boom in immunological research. A newspaper story of David, the immunodeficient child who emerged from his bubble at the age of twelve to be kissed by his mother for the first time in his life, carried the casual comment that doctors gave him a bone marrow transplant which "had been treated to reduce the possibility of rejection." Implicit in that statement is an enormous background of research in basic immunology.

Because the vertebrate immune system is a long way from being completely understood, there will continue to be opportunities in industry for immunologists. The best-known branch of immunology is vaccine development. But a young immunologist seeking a post in industry also needs to realize the extent of such development and its economic potential. The industry has moved from "standard" immunizing injections, using killed or attenuated organisms, into interferon production, immunizing antigens produced from cloned genes, and diagnostic reagents that rely on monoclonal antibodies. In addition to the demand for human vaccines and immunodiagnostic methods, there is an enormous need for veterinary pharmaceuticals, including those developed and manufactured by the same techniques used for the human products.

While the intellectual contributions of immunology have definite medical and economic importance, immune function is in itself one of the more intricate biological processes. If you are the kind of person who can see through this intricacy to detect new ways of solving problems, then you may find a comfortable place for yourself in industry.

It would be easy in a book such as this to go on for pages about specific opportunities in industry. What would finally emerge from such a consideration would be something any economics major could tell you: American business is significantly more varied than the average young person realizes, and if your training has money-making potential, there is a job waiting. For our purposes, the final question will be: Can I take this job and remain a biologist? The critical issue is the nature of the company that offers you a job. Does it want your insight, your determination to solve a problem, or only your ability to earn it money in the short run? This question is directly related to another: Do I want to remain a biologist or become a worker? If your answer is the former, then you must seek that position in industry that provides some minimum amount of the intellectual freedom characteristic of academia. We are, logically enough, back to the difference between being and doing.

4.

The options for a biologist who wants to work in government are as varied as those in business. Examples of government biologists range from a beleaguered refuge manager under the administration of a James Watt to full-time researchers at the National Institutes of Health (NIH), some of the most privileged scientists on Earth. The sunburned towhead who hauls in a gill net for state game and parks will, if asked, call himself a biologist. So will a Washington bureaucrat, a desk-bound ecologist in the Environmental Protection Agency (EPA), an Alaskan game ranger, a curator in the Smithsonian, and a world-renowned parasitologist checking calf feces at the U.S. Department of Agriculture (USDA) lab near Beltsville, Maryland. The military is filled with people who call themselves biologists. All are making a living through the knowledge and understanding of living organisms.

Opportunities for biologists in government arise from a perceived need for Washington, as well as states, counties, municipalities, and districts (for example utility districts), to assume certain functions that individuals cannot and corporations will not. These functions include maintaining collections of items of overall cultural significance (museums), evaluating goods intended for public consumption (EPA, Food and Drug Administration or FDA, USDA), solving general or potential health problems (Public Health Service or PHS, NIH, National Cancer Institute), administering public resources (Department of Interior or DI, state game and parks and city parks and recreation departments), and upholding and strengthening the national ideology (Department of Defense, or DOD).

Career opportunities in government also vary not only with the economic times and the state of technological development, as in business, but also from conditions that derive primarily from government function—that is, the political and military climates. The Reagan presidency is an example of an administration notoriously hostile to the preservationist movement, and just as notoriously given to showmanship economics in which an image of fiscal restraint was maintained at the expense of agencies such as EPA. Although they are alarming, such political stances usually have only a short-term influence on employment opportunities: It is characteristic of nearly all levels of government that a large fraction of its employees survive significant fluctuations in political climates. (Some critics find this degree of job security one of the problems with government employment and a major source of the lethargy with which government often seems to operate.) These comments do not apply, of course, to political appointees.

The military has historically been an excellent source of employment for biologists, who sometimes masquerade as physicians. Perceived national interests often stimulate government activities that are discovered, belatedly, to require biological expertise, perhaps the most notable example being the case of Walter Reed and the Panama Canal. His victory over yellow fever is a classic tale of tropical medicine, a field which is often

characterized as much by basic biology as by our present image
of the physician. (In the tropics, health-care professionals are
required to solve big problems involving snails and worms as
often as they prescribe tranquilizers.) With World War II
came our modern awareness of the logistical importance of
certain kinds of biology. War is great for parasitic protozoa and
worms, fleas, and lice, as well as viral, rickettsial, and bacterial
diseases (see Zinsser, *Rats, Lice and History*). Experience has
shown that wars have resulted in as much contact between
humans and nature as between humans and humans. In fact,
few wars have been fought in which more soldiers fell to
hostilities than to disease.

The military is vitally concerned with biology because
outside the boundaries of the United States, disease transmission is often a complex problem involving several kinds of
organisms. Malaria, one of the most prevalent infectious diseases on Earth, depends on an interaction of at least three
species: the mosquito vector, the parasite itself, and the human.
Schistosomiasis, easily as widespread as malaria, similarly involves the parasite, snails, and humans. The dynamics by
which these organisms interact may vary locally, and armies
can quickly become ready and immunologically naive partici-
pants in such interactions. The result can be devastating; a
soldier flat on his back with malaria is as ineffective as one
felled by hostile fire. Commanders who have ignored this elementary fact have always paid the consequences.

Both the United States Department of Agriculture and the
National Institutes of Health employ biologists in significant
numbers. These agencies are notable in that they conduct intramural research—that is, research done by agency employees
—in addition to their sponsorship of extramural projects
through grants and contracts. The most prestigious intramural
positions are typically held by people who've established themselves as productive scientists early in their careers. They often

do basic research in epizootiology, entomology, immunology, biochemistry, pathology, nutrition, neurophysiology, and cellular and molecular biology. Although they are likely to be working on organisms or problems that seem to be of practical importance, the actual research is often so basic that its precise application cannot be foreseen.

You will probably not be competitive for one of these positions unless you establish a significant reputation, through publication, either during your post-doctoral work or within a very few years thereafter. Such accomplishments may seem an impossible task to an undergraduate, for it is often exceedingly difficult for a student to visualize himself or herself as a successful scientist. Many professionals, however, look back on their own lives and see themselves as fairly typical undergraduates. Productive careers are much more the result of being able to match interests and abilities with opportunities than of having supreme intelligence.

The Department of Interior, another large employer of biologists, hires scientists of a different breed from those of NIH, USDA, and PHS. DI biologists most often have training in ecology or wildlife management, and tend to have worked their way up through the ranks from a position like park naturalist. DI biologists are, for example, refuge managers, wildlife biologists, and foresters; their duties range from administration of a bird banding program to research on tree growth. The advantages of such employment are obvious to a person who loves to be alone. Conversely, for an urbanite, a three-year stint as manager of the Crescent Lake National Wildlife Refuge twenty-six miles down a dirt road from Oshkosh, Nebraska, has little to recommend it. For people who require regular social contact, there are not enough avocets, phalaropes, pelicans, western grebes, or dollars on Earth to offset the loneliness of such a place. Some thrive on isolation, however; their communications are with their ideas; there is interchange between the throbbing complexity of a marsh and that part of the soul that needs to speak and listen. Again we are at the point where a book can only tell you to see yourself as you really are, then seek a place where you fit. But if you take a wife or husband to Crescent Lake, be certain that he or she can also trade humans for curlews and never bat an eye.

5.

A museum is one of the most fascinating places in which humans find themselves, either as employees or visitors. In addition to public programs, major museums usually support research activities and collections which are used not only by resident curators but also by visiting scholars from around the world. The position of curator requires a person to be at home with rooms full of preserved and fossilized insects, ticks, mites, snails, clams, fish, reptiles, birds, rodents, flowering plants, lichens, fungi, ferns, parasites, and a host of additional kinds of specimens. In this respect a museum curator is similar to a refuge manager, in that extensive and continuous contact with other humans will interfere with his or her work. The primary responsibilities of curators include maintenance of and research upon the museum collections. Included in their specific duties are record-keeping, identification, field trips, and the lending and borrowing of specimens.

It is true that a person can become buried in his or her collections. On the other hand, some of my strongest memories are of G. M. Sutton suddenly walking to a cabinet, pulling open a drawer, and giving an impromptu lecture on the variation revealed in a row of beautifully prepared skins. It is a rare experience for a student to have a curator actually explain what a collection reveals. Suddenly the concepts of geographic variation, development, and speciation come alive even though the organisms are dead. This kind of experience underscores the fact that museums are our repository for the evidence to support the naturalist's world view.

In many respects, museum curators are as privileged a class as research scientists at the National Institutes of Health. I recently had the opportunity to review the activities of several curators at a major state university museum. A year in the life of any one of them would sound like purified excitement to the average citizen. For example, the entomologist was responsible for the records, fumigation, labeling, and alcohol changes in a collection of 1,750,000 specimens worth an estimated minimum of twenty dollars each. During the year he acquired an additional three thousand specimens; hosted nineteen faculty scientists, mostly from other universities, and fifty-one graduate and undergraduate students who used the collections for research; identified thirty-two specimens for scientists in ten states and five foreign countries; taught a four-credit course; went on field trips to central Mexico and western Nebraska; visited museums in Ohio, Berlin, Paris, and London to study their collections; had seven nonpaying consulting jobs; published four papers; attended four scientific meetings; presented a talk at an international congress in Germany; served as president of an international society; and submitted eight grant applications, five of which were funded. In that same year, the state appropriated a total of six hundred dollars to support his activities, evidence that the world has yet to understand the importance of keeping a record of what insect species are present on Earth.

Museum biologists may also find themselves involved in exhibits preparation and public relations. While artists and craftsmen usually design and construct exhibits, the curators often supply much of the biological expertise behind the show. Of all the activities in which I have observed biologists at work, the preparation of museum exhibits has seemed to be the most like play. I have felt as if these curators, after years of study derived from deep love of a particular group of organisms, finally get the chance to show off what they like best to people who have come to the museum in the first place because they were hungry for entertainment more substantive than a ball game.

6.

The final major area in which biologists make their living is what I will call, for lack of a better term, the private sector. Our attention now turns from organizations that want your talent, to talent that wants a place for itself badly enough to generate one. This is the age-old story of entrepreneurship. It is the heart of the American Dream. There are not many places in the world where more opportunity exists than in the United States for a person, even a biologist, to lay his or her talents on the line. Of course this is no path for those who are afraid of failure.

Few institutions are more American, or more private, than the consultant. Consultants prosper because of three factors: (1) a business or government need for information upon which to make decisions, (2) the intellectual insecurity of organizations that rely on information, and (3) the cost of obtaining the information. In the history of business, successful consultants have often been those who convinced the consumer of a real need for information, thus combining the role of promoter with that of the informer. In some consulting areas—for example, environmental impact—opportunity fluctuates with government policy, while in other fields, such as molecular biology investment, it follows business trends. For a biologist seeking sustenance in this portion of the private sector, the questions to answer are: What kind of biological information does industry or government need? Can I supply it more cheaply or convincingly than they can themselves? Will they pay enough for it so that I can make a living?

On a more organized level of the private sector are the National Audubon Society, the Nature Conservancy, and the Sierra Club, all examples of institutions that survive because of a sense of mission among those with common biological interests. The support for these groups comes mainly from private sources, although magazine subscriptions and mail order operations also contribute to their incomes. Staff positions that may be obtainable range from Audubon Society camp and tour leaders to Nature Conservancy land stewards.

Also in the private sector are the commercial nature tour operators, wildlife artists, writers, photographers, and film makers. If you have a big enough boat and access to sea bird colonies or whale breeding grounds, you can probably make a living. Your ability to compete in the nature tour business may depend on the extent of your knowledge of the organisms your customers are paying to see. The financial resources of that segment of the population interested in biology as recreation are sometimes a surprise. An ornithologist colleague of mine recently accepted an all-expense-paid six weeks in India as a tour guide for a group of bird watchers. If you calculate the cost of this venture, then count the number of companies advertising such tours in your latest issue of *Audubon,* then estimate the amount of business they must do to stay solvent, you'll have a rough idea of the magnitude of this audience. As is true of any business that depends on consumers' discretionary income, current expenditures are likely to be but a small fraction of the potential resources that could be diverted into private biological education and entertainment. All you have to do is figure out a way to get at them.

Artists, including painters, sculptors, film makers, photographers, and writers, are notoriously poor. With few exceptions, those whose subjects are biology are no different from the rest. A common question I was asked when my first book, *Keith County Journal,* was rave-reviewed in *Time* magazine was: Are you going to New York to become a full-time writer?

The thought of doing something quite that stupid had never entered my mind. But dreams of the writer's freedom had, and were only reinforced by the public recognition that came with book authorship.

All such thoughts were put into perspective a few years later when I spent some pleasurable days with a former fellow graduate student. This man had received his M.S. from the University of Oklahoma, his Ph.D. from the University of California at Davis, had done post-doctoral research at the University of Colorado Alpine Research Institute, and had found employment at a state university, where he was ultimately denied tenure. His response was to return to his beloved mountains to become a nature photographer. Ten years later we met on the prairies for an unsuccessful joint venture. I commented that he and his wife never looked healthier, that they were, after all, alive and free while my "success" at a large university had brought an administrative position and a daily dose of high blood pressure medication. At a local pub I waxed philosophical and venturesome on the subject of telling the system goodbye and retiring to the wilderness to become a nature writer and artist.

"Don't jump into it," said his wife with an unusual amount of authority.

I dropped the subject. But I did buy two of his photographs and my wife bought another. The pictures are haunting; they give me a view of life that suggests an organism's chance to fail, or succeed, on its own ability. Is this photographer a biologist, or is he an artist? He is using a deep knowledge and sense of biological material to produce one version of the truth. His technical skills are used in combination with that knowledge, and with his sense of what is important. His product is consumed by those who are interested in its message. The ultimate value of what he is doing is not easily assessed in the short run. He has to raise money to support his work. He derives pleasure and a feeling of accomplishment from his exploration of nature. He solves problems. His values are those of the scholar. Most of these things could also be said of any scientist.

I'm not sure what this tells us about becoming a biologist. It may suggest that contact with the world of organisms induces a touch of the artist in a person, no matter how technical things may seem on the surface. But my friend's wife's advice will not go away. It continually speaks of the trade-off between making a living and doing original and creative work in any field, be it art, writing, or biological research. As I look over all the people who would call themselves biologists, and who are making a living at it, my friend with the camera seems to fall within their range of types. At one extreme, some call themselves biologists and the label easily sticks because of some requirement of their jobs, no matter what is in their minds. At the other extreme are those who upon serious consideration are revealed to be thinking like biologists, yet don't seem to be making much of a living at it.

It is perhaps in this analysis of the private sector that history and biography are of the most use to us. History shows that there are few things more valuable than an individual idea, and that certain ideas can be converted into money, power, good, evil, comfort, security, misery. History also shows that a population of humans will, given the least opportunity, produce individuals who can change the course of history with their ideas. Biography, on the other hand, suggests that the private lives of such individuals are anything but smooth. Such people have rarely made much of a living. Even more rarely have they pursued formal education for the purpose of certification and employment. In fact, a significant fraction of them reveal a record of conflict with formal education, evaluation, and endorsement systems, an experience well known to highly creative people.

My advice, if you are beginning to see yourself in the above paragraph, is to invest some more time in determining whether you are indeed a highly creative person. If the answer is yes, then you are faced with two general alternatives: suppress that creativity long enough to get certified and employed, or let it drive you where it will and then define the results as success.

7.

You may have found yourself a comfortable slot in one of the areas discussed in this chapter, and may be making a comfortable living, but is it biology? This is the critical question the purists will ask. There is no correct answer that does not invoke the values and world views discussed earlier. Humans are notorious for their ability to modify, manipulate, and influence the nature of phenomena, including jobs. It is difficult to imagine a true biologist—that is, one in whose veins the naturalist tradition flows strongly—who sees the world as do my two student friends in the pizza parlor, and who holds values George Sutton could have held, who would not be a biologist no matter what he or she was doing for a living. The most likely outcome is that this person would, through his or her values, views, and approach, simply convert a job into biology in addition to whatever else it might be.

I have seen this matter of employment, or a career—and they are not necessarily the same—become an obsession with students as well as their parents. I have also had the experience of, within a twelve-hour period, responding to a common student question ("What can you do with a degree in biological sciences?") and seeing a stunning performance of the Hubbard Street Dance Company from Chicago. These two seemingly disparate events stick together in my mind because in the back of the HSDC program was the statement that Lou Conte, Artistic Director and choreographer, was a zoology major at Southern Illinois University. This curious bit of information brought back memories of graduate students who finished their degrees in the midst of the worst job market in decades, not having any idea of what they were going to do, who struggled in various temporary appointments but who are now tenured professors at major universities. It also reminded me of friends who went to the finest schools, worked with famous scientists, landed enviable positions in government research labs, but who are now attorneys instead of biologists.

"You can't predict the future" seems a trivial cliché as well as an unsatisfactory answer for someone who wants to know what opportunities there are for biologists. In retrospect, those I've known who really wanted to be biologists now are, and some of those who didn't care have gone on to other things. Perhaps this is a good time to relay the manner in which my own teachers handled the question of my future (see Acknowledgments). Harley Brown never spoke of work; all life was pleasure, or if not pleasure then at least interesting. J. Teague Self assumed his students would be successful, and because he did, so did we and didn't worry much about it. Leslie Stauber never said so directly, but I always felt his only concern was that I was willing to go to Nebraska. And Sutton looked me straight in the eye and said, "There will always be a place for a good person!"

5.

RESPONSIBILITIES

> "Madam," he asked, "how do you people talk to your animals?"
>
> "Do you not talk to yours?"
>
> "No."
>
> "You stay with them. You watch them. You put your hands on them and feel how they feel. You look into their eyes. You listen to the tones of their cries and their calling to one another. You make sure that when they begin to understand that you understand them, you do not miss the first tones of what they say to you."
>
> —from DORIS LESSING, *The Marriages Between Zones Two, Three and Four*

1.

Does a scientist have responsibilities? Yes, and they extend beyond the obvious ones such as ethical behavior and objectivity in research. In this final chapter I'm going to try to delve into some areas of responsibility that the average student may not have thought of, or at least not considered very seriously. My analysis is philosophical; not all will agree with it. But there should be no question of the fact that educated citizens do have, by virtue of that education, certain obligations. If they are in disagreement with the ones suggested by me or others, then their first responsibility should be to formulate a list of their own duties to themselves, to their profession, to society, and to the planet that supports them.

A responsibility is an obligation, a requirement that one accomplish or control certain things, or at least attempt to do so. For example, the principal investigator on a research grant is the one who ultimately must see that supplies are ordered, equipment is maintained, experiments are conducted, and expenditures limited to the available funds. Obligations can also bind individuals into groups, as when a department must offer introductory biology for nonmajors in order to satisfy the wishes of a College of Arts and Sciences faculty. Often a feeling of obligation can be as effective as an obligation itself in motivating behavior, as happens when a scientist feels a need to introduce ideas into a discussion regardless of whether he or she was asked to contribute. The responsibilities I have just described are ones that function to determine, in part, the behavior of professional biologists. But rather than analyze this phenomenon, I would prefer to ask: Is there behavior that should accompany the role of biologist in the last part of the twentieth century? My answer is yes, and it is derived from the well-publicized and well-founded concern biologists have for the future of our species.

A scientist's responsibilities rest on his or her possession of knowledge that nonscientists do not have. An excellent example of this kind of knowledge is the information about plant physiology that supports predictions of the ecological effects of acid rain. Botanists have little trouble believing that acid rain can eventually destroy not only our forests but also the productivity of our streams and rivers, thus altering vast segments of the nation's economy, while a less-educated citizen is sometimes willing to dismiss such predictions as unproven speculation. Compounding this problem is the fact that scientists, being human, do not always agree with one another, especially on matters of interpretation and prediction. The general public is not very patient with academic discussions. When one scientist predicts the deforestation of America due to acid rain in fifty years, while another gives us two hundred, and a third says the issue will be resolved in other ways (nuclear war, end of fossil-fuel supplies), the man on the street often shrugs and turns to the sports page. This latter act does not make acid rain go away. Instead, it makes the scientist's job of public education more difficult.

A scientist has an obligation to share knowledge because all individuals can act in ways that affect the common circumstance. Scientists are almost unique in knowing a certain "real world"—that is, the world of physical reality. They are responsible for sharing this knowledge because physical reality is often utilitarian, providing its users with power over fellow humans, or power to alter our common environment on a grand scale. Those who are potentially subject to environmental deterioration or the use of power need information with which to counter these forces. Underlying the scientist's obligation to disseminate knowledge is the fact that our society depends on sophisticated technology. My contention is that we are all better off when citizens, as well as political leaders, know the nature and limits of our technical abilities.

Purists will argue that scientists should be above politics. But history shows that politicians are not above using scientific discoveries. It is the fundamental nature of humans to manipulate, experiment, explore. This nature, however, is sometimes manifested as brazen disregard for the intent of discoveries. The best example of such an instance is the development of nuclear weapons. Freeman Dyson eloquently describes this process, and the fate of the scientists involved, in his book *Disturbing the Universe:*

> Oppenheimer was driven to build atomic bombs by the fear that if he did not seize this power, Hitler would seize it first. Teller was driven to build hydrogen bombs by the fear that Stalin

would use this power to rule the world. . . . But each of them, having achieved his technical objective, wanted more. Each of them became convinced that he must have political power to ensure that the direction of the enterprise he had created should not fall into hands that he considered irresponsible.

Although Dyson was recounting his career as a physicist and reflecting on the significance of events he'd witnessed when he wrote those words, they nevertheless tell us much about the ways we use, and perhaps come to be used by, our discoveries. *Disturbing the Universe* begins with Dyson's memories of a children's book, *The Magic City*, by Edith Nesbit. In the city of the story, the hero Philip can have anything he wishes, but if he wishes for a machine, he must keep it for life. At the end of his career, Freeman Dyson saw the symbolism in this tale and used it as a framework to tell his own story of the nuclear age. I am inclined to believe the Cambridge, Massachusetts, city council also understood Philip's dilemma when it passed a city ordinance against gene-splicing experiments. One person cannot control forever a discovery that is powerful enough to become a tool for government or industry. This is the reason why the public as a whole needs scientific knowledge.

There are those who feel the best way to prevent the misuse of scientific discoveries is to inhibit research—that is, to restrict inquiry. These people ignore, sometimes intentionally, the contribution that science has made to our lives. This contribution is more than economic, it is also philosophical. Science, like art, continually shows us that of which we are capable, and it demonstrates unequivocally that we have a long way to go before we reach our intellectual limits. By extension we must question whether we have reached our limits in other ways. There are those who believe that hunger and oppression are integral parts of the human condition. The scientist responds by saying that humans once believed a flat Earth was the center of the universe.

Thus the role of scientist carries two major integral responsibilities: (1) to speak the scientific truth, as we know it today, to the general public, and (2) to demonstrate through visible actions that skepticism, curiosity, and questioning are natural, and indeed expected, human behaviors. The discharge of these responsibilities involves speaking out on local and national issues whose resolution requires the use of scientific information. The best known example of this in biological history is Rachel Carson's book *Silent Spring*. Although it was published in 1962, it still provides a powerful reading experience, made all the more so by our knowledge of what it accomplished—namely, the creation of a national awareness of the danger of pesticides, culminating with a ban on the sale of DDT. Today, sociologists who make a study of trends, and the factors that produce them, view *Silent Spring*, along with Ralph Nader's *Unsafe at Any Speed*, as examples of the general phenomenon in which a single published work alters public opinion. The world is no less in need of statements like *Silent Spring* now than it was in 1962.

It is important for the public to view the scientist as a person who not only seeks the truth through observation and interpretation but also continually searches for new observations and welcomes discoveries. This perception is important because science is the keeper of a belief system in which reality is established by observation as well as some formal thought processes (for example, logic) that in turn establish relationships and predictions. Material accomplishments such as flights to the moon remind us that the belief system of science does indeed produce power. There are other belief systems, such as those in which truth is established on the basis of statements from a single individual, that when adopted by nations typically result in loss of intellectual, and subsequently political, freedom. I am not implying that we can remain free only through use of the scientific method. I am implying that when truth is based on impartial evaluation of evidence, as it should be in science as well as some other human affairs, then our chances for full development of human potential are the greatest.

Scientists also have an obligation to continually demonstrate that science itself is largely neutral. The past use, or misuse, of scientific discoveries should be laid, by articulate scientists, squarely on the shoulders of those responsible, who are very often the politicians. The neutrality of inquiry must remain sacred. Without neutrality, there is no true scientific inquiry, and perhaps no true inquiry at all. We have, at least in theory, the option of deciding that there will be no inquiry, that the discontinuation of unbiased investigation is one way to halt threatening discoveries. But it is questionable whether humans will ever stop searching for explanations of the things they see. Our driving inquisitiveness may be so much a part of even our genetic makeup that we are unable to suppress it for long. The use of discoveries, however, involves moral choices. So long as individual scientists remain aloof from moral issues, discoveries will continue to be used by people who do not understand their full implications and power.

Finally, because science is an activity that produces change, it is by its very nature disruptive. Individuals do not usually welcome disruptions, especially ones not of their own choosing. However, the list of ways that science has changed our world and overturned our fundamental beliefs could go on for volumes. By its length and breadth alone, this list would place seemingly insoluble human problems into perspective. For example, Native American populations were devastated by smallpox during the nineteenth century, and the plague epidemics of the fourteenth century equally reduced the population of Europe. At the time, there were no known solutions to such problems. Today, we know the life cycle and control methods for *Pasteurella pestis*, the causative organism of bubonic plague, and the World Health Organization claims the smallpox virus is extinct. These historical examples suggest that to modern science, problems that defy solution are viewed as problems that have not been attacked with the right resources. By "resources" I mean ideas as much as time and money. Thus the scientist stands as a constant reminder that things are not fixed, and that the human mind does have powers that maybe are yet to be tapped. It is the responsibility of the scientist to accept this role and to fulfill it for a society that is ever in disagreement over what it wants changed and what it wants to remain fixed.

2.

Are there responsibilities that a biologist has that other kinds of scientists perhaps do not? Yes, and these obligations are derived from a biologist's specific understanding of living, versus nonliving, systems. Today, to define a biologist as one who understands living, as opposed to nonliving, systems is to bring oneself to the disreputable edge of vitalism, a doctrine that asserts life cannot be explained by the laws of chemistry and physics, but instead depends on some "vital principle." Vitalism is anathema to reductionist science, which seeks explanations for biological phenomena that are entirely consistent with chemistry and physics, and in the process assumes that all living systems (organisms) can potentially be described by the behavior of nonliving (chemical and physical) ones. I am not going to cross the boundary into vitalism in this book, but will approach it more closely than would some others.

The philosophical problem faced by a modern biologist is this: Although analytical techniques, including formal mental ones, have given us a highly detailed knowledge, down to the biochemical level, of a number of organisms (*Escherichia coli*, *Paramecium* species, strains of inbred mice), it has yet to be demonstrated that this knowledge logically explains all aspects of the organisms' lives. For example, let us ask a question: If we knew the entire DNA sequence of a sandpiper, had an inventory of its chemical compounds, knew how the expression of its every gene was regulated, could describe its enzymatic reactions, and knew the structure of all its proteins, could we then logically explain, from that information, the bird's annual migration from Canada to Argentina, or nesting behavior so tender it brings tears to the eyes of a seasoned field man? If your answer is yes, you are operating mentally in the philosophical tradition of a committed reductionist. If your answer is no, you have one foot in the camp of the vitalists.

These alternative answers to the above question are important because they influence the manner in which you, as a biologist, will present your work to the world. A reductionist's approach will have you constantly studying parts of systems, assuming your discoveries have implications beyond their immediate context, and describing your research in terms of isolated functions. The opposite approach will have you occupied with events and processes that depend on the organism's integrity—that is, upon the sum of its parts. I suggest that on the basis of our present understanding, perhaps we should not be so quick to shun completely a modern vitalism. The public is more easily captivated by the beauty and behavior of whole organisms than by the chemical reactions that power them. Furthermore, the intact plant, animal, or microbe provides a contextual framework within which lay persons can organize, and gain perspective on, basic life processes. This point is obvious to anyone who has taught freshman biology; oxidative phosphorylation does not seem so meaningless when it is shown to occur within mitochondria that are inside muscle cells that contract to close an ant's jaws around an outlandishly large piece of food.

A modern vitalism might be expressed as follows: In the sandpiper example, migratory and nesting behavior are properties derived from the arrangements and relationships of the bird's component parts, but we have no idea how organization *per se* results in the wondrous activities we observe. It is this kind of vitalism that contributes heavily to the context within which a true biologist thinks and acts. Thus the biologist has, by virtue of his or her interests, the obligation to continually attempt (1) an integration of parts into a whole, and (2) an explanation of the whole in which both the behavior of the whole, and the role of the part, are considered. This manner of thinking is, or at least should be, characteristic of one who considers the function of an organelle relative to the life of a cell, of a cell relative to the life of a tissue, and so forth up to and including the roles of whole organisms in the organization of an ecosystem. With this kind of perspective, an average citizen should be able to metaphorically place his or her time on Earth into a context that includes the entire planet and its evolutionary history. A biologist has an obligation to explain, and perhaps promote acceptance of, this metaphor.

Why does such responsibility fall upon the biologist? There are two reasons. First, because organisms are by definition complex systems, a person who studies these, or parts of them, must always bear this complexity in mind for the sake of his or her own intellectual development. There is, in that respect, an obligation to oneself. But secondly, the biologist exemplifies for the public a mind-set that it trusts to make integrated observations. As a role model for society, the biologist above every other kind of scientist should demonstrate the futility of searching for simplistic and purposeful answers to complex natural problems.

Within the confines of the profession, a biologist can get by with a narrow focus and a failure to attempt to integrate isolated observations into a whole. While as individuals some of us can shirk our intellectual responsibility, be assured that others will gladly accept it, and will be thought of as better biologists for having done so. Outside the confines of the profession, however, the biologist who denies the complexity of systems is failing to fulfill his or her function for a society that was never in more need of an example.

3.

Does a biologist have a responsibility to the human species? This is an especially interesting question. Any attempt to answer it forces us into a consideration of our species' characteristics (as opposed to the taxonomic characters, but more in line with our sociobiological traits than some elusive "human nature"). What have we shown ourselves to be? Do we believe, as a group, what we have shown ourselves to be? Does the biologist have anything to contribute to our vision of ourselves? If so, does such a contribution bring us to some moral, ethical, philosophical, and subjective aspects of life that are usually considered outside the domain of scientists? These questions are all related to my first question, and I will suggest, in the order they are asked, their answers.

Humans have shown themselves to be animals, members of the order Primates, family Hominidae, relatives of the gibbon, orangutan, gorilla, and chimpanzee. As for the second question, collectively we do not readily accept these relatives as legitimate ones; that is, human societies, regardless of how apishly they may behave, do not act as if they believe what biology has revealed to us about ourselves. In answer to the third question, biologists have also shown that we possess immense powers, traceable primarily to the brain and its as yet ill-defined derivative, the mind, but complemented by certain physical attributes, most notably the hand (see Melvin Konner's *The Tangled Wing*). Our nearest relatives do not, so far as we know, have all our mental abilities. This suggests we are left with two major possibilities as to the origin of our intellect: (1) Divine Provenance, and/or (2) the unique physical structure of the human brain. The argument between those who favor one source over another can be reduced to triviality by simple theological mathematics: Brain structure and mind function may be defined as God's works, or, God may be said to have directed the evolution of whatever brain structure gives rise to higher intellectual powers. Alternatively, of course, brain structure and mind function could both have evolved without help or interference from Omnipotence. Regardless of the origin of our equipment, we have it, use it, and live in a world that is shaped by it.

The reality of our world as we have shaped it is fairly stunning: With the current nuclear arsenals in place, human thoughts, those untraceable and intangible products of the mind, can end life as we know it on the only planet in the universe known to support it. If we credit Einstein with responsibility for our entry into the nuclear age, then our situation is all the more stunning: An individual, an employee in a Swiss patent office, who never did what a modern molecular biologist would call real lab research, gave humanity thoughts and ideas that it fiddled with, in the end converting them into the potential for life-destroying war.

Finally, does biology's contribution to humanity's self-image bring us to philosophical, ethical, and moral considerations that are usually thought to be outside the domain of science? Attempts to answer this question have consumed endless hours of barroom talk; there is no reason for the discussion not to be aired more widely. The human experience includes an exceptional range of activity (as suggested before in references to LeShan and Margenau's *Einstein's Space and Van Gogh's Sky*), of which biological science as it is usually practiced occupies only a small fraction. But every biologist understands the sense in which we are as much a part of nature as a bumblebee. Therefore our profession ultimately must address the problems of humanity or try to answer the potentially embarrassing question: Will those who profess to seek answers to living riddles shy away from the most puzzling species of all? Is one who calls himself or herself a biologist in default if a single species' attributes are left to anthropologists, sociologists, philosophers, and writers? I think so. Our responsibility to the species is to study our biggest problem: Ourselves.

In this regard, the Cavalli-Sforza and Feldman synthesis (described in *Cultural Transmission and Evolution: A Quantitative Approach*) is perhaps of profound significance. Their demonstration that the distributions and movements of rumors, innovations, mutant alleles, and infectious diseases in populations are analogous (that is, they can be described by the same mathematical models) in effect equates these entities in some disturbing ways. With devil's advocate strictness we must say that similarities in behavior do not necessarily prove similarities of fundamental process. But analogy can nevertheless serve to direct our thinking along lines that we may not have previously tried. What Cavalli-Sforza and Feldman have told us is that our mental products can and probably should be considered as valid an example of biological forms as viruses or parasitic worms. With that idea we proceed to the logical evolutionary question: Are the species' thoughts interacting with a changing environment to produce an evolving population of concepts? The authors make a convincing case that the answer to this question is yes.

Furthermore, the environment in which these expressions live is largely one of their own construction. Culture is, in effect, a complex self-feeding system that can direct its own evolution. Our mental powers have made possible an "escape" from the "restrictions" of the natural world, those restrictions being boundaries imposed by the range of forms, colors, sounds, situations, functions, and behaviors of the nonhuman environment. Admittedly, this range of boundaries is exceedingly large, but in the final analysis, it is finite. The human mind, in contrast, knows no limits. It is free to construct any environment, including imaginary and surreal ones, and, most importantly, to suspend the rules that control "natural systems" in order to build a place in which to operate. There is no more obvious example of such mental constructs than the ubiquitous multitude of science fantasy pulp paperbacks in your local drugstore. But mental abilities and inclinations can also function to delude us in potentially dangerous ways. For example, no matter how much we believe we can escape our ties to the life processes that operate on Earth, this belief is founded in self-deception.

The biologist therefore has an obligation to our species to explain the nature of the somewhat naive belief in cultural escape. The argument for this obligation is a very biological and scientific one. It assumes our species has an inherent value that is greater than that of other species. We possess this higher value because we are the only species to have demonstrated our ability to explore the universe beyond our ecological niche without having to adapt to that environment through organic evolution. My conclusion—namely, that this ability makes us more important than even sea otters—is founded on a belief in the value of knowledge itself.

Does this assumption of high and unique value for the human mind necessarily justify unconscious exploitive and short-term, self-serving behavior relative to our planet? No, and in fact, it should do exactly the opposite. The scientific community is the repository for society's noncommercial interests in the universe. Within the biologist community is housed the species' basic curiosity about other organisms and about the processes that make life itself possible. It is in the vested interest of our species that this noncommercial interest survive; for it to do so we must maintain an environment in which the brain can continue to support the mind's explorations. However, humanity often seems to behave counter to its own welfare, and fundamental principles that every biologist knows from studying bacterial growth in closed systems (growth cannot be maintained indefinitely) are being ignored by the collective mind. The mind may be distinct from the brain in some regards, but no mind can exist without plasma membranes, sodium pumps, ATPase, DNA, transcription, translation, and protein assembly. There is every reason to suspect, furthermore, that the mind, as distinct from the human brain, may also require the planetary grandeur symbolized by the sea otter, whooping crane, and California condor. These last statements, so obvious to even a beginning biologist, seem to have made no significant impression on the world's population. A biologist has little choice today but to take upon himself or herself the task of correcting this global self-deception.

4.

Is culture a fit subject for biologists? Traditionally, biologists in their formal work have avoided the humanities and social sciences. Those who have ventured into these areas (for example, the sociobiologists) have found themselves in the midst of controversy. They have also been criticized for sloppiness by those accustomed to the analytical comfort of biochemical methodologies. But in response to this point I have repeatedly suggested that the biologist does have legitimate responsibilities outside of pure science which may, if accepted, erode the sequestered comfort of the laboratory. These responsibilities encompass domains of human activity we usually include within the concept of culture.

The biologist encounters culture nowhere more directly than in the introductory classroom. Large-lecture format biology for nonmajors is a concentrated microcosm of society's range of values, attitudes, beliefs, economic conditions, and aspirations. What, then, is the responsibility of a professional biologist toward the introductory student? The foremost task should be that of altering, if necessary, mind-sets and thought processes so that these people:

1. want to consider the long-term consequences of their actions,
2. are able to interpret most new material (for example, that from the media) into an existing context which includes the nonhuman living world,

3. have some sense of the processes that operate in living systems at many levels,
4. look for evidence of the true character of things by studying their ontogeny, and
5. are able to place *H. sapiens* into a proper time/space/species diversity framework—that is, to understand that the world did not begin in 1965, that it may not even be unique, and that we are certainly not alone in it.

This last is a special challenge. For example, although regularly publicized, new hominid fossil discoveries do not necessarily seem to add depth to the average citizen's picture of himself or herself. Neither, if political activity is the true manifestation of the collective consciousness, do reported studies of phenomena such as acid rain. It is up to the biologist to expand our self-image beyond our immediate circumstances. In an average large state university, a biologist walks into an auditorium in late August and stands in front of three hundred freshmen. If the responsibility is to be fulfilled, then four months later the three hundred must walk away aware of themselves as individuals, as part of a large and complex, yet closed, environment, and as a small yet significant part of a four-billion-year progression of life forms.

I contend that a straightforward presentation of basic biology does not accomplish these ends. Merely understanding the processes of gene expression will not convert an average freshman into the kind of thinking and feeling person just described. Nor is there anything in cell ultrastructure, no matter how beautiful, nor in nutrient cycles, or inheritance, anatomy, taxonomy, or biogeography, that will in itself carry out the teacher's obligation. It is the responsibility of the biologist to make biology a cultural phenomenon for this semester's model of humanity. It is almost impossible to accomplish this feat without some understanding of the social sciences

and humanities, as well as a willingness to use that understanding in the classroom. Humans do not generally act, either independently or in groups, as scientifically rational bod-
ies. Instead, their actions spring primarily from feeling,
purpose, or a psychological history that shapes responses to
stimuli. It is doubtful any biologist can affect the latter; it is
the responsibility of all biologists to try to change sensibilities
and motives.

At the scholarly level, culture should also be an appropriate
subject area of research for the biologist, if not at present, then
certainly in the near future. I had the pleasure not long ago of
passing an afternoon in the company of my biologist colleagues at this university. The daily cares of classroom, committee, and research were set aside and we asked ourselves to
think. In the safety of a conference room, no anonymous outside reviewers to cast aspersions on our most far-reaching
ideas, we addressed the question: Where will the forefronts of
biology be in the year 2004? Included in the group was a
member of the National Academy of Sciences, cell and molecular biologists, my herpetologist friend from earlier pages, several ecologists, botanists, animal parasitologists, teachers, and
administrators. Some conclusions were interesting: Molecular
biology will be dead; the complete DNA sequences of several
multicellular eukaryotes will be known and stored as memory
banks; some of today's serious subdisciplines will be practiced
as hobbies if at all. Other conclusions were more humble:
Biology will still be facing the unsolved problem of how genetic information is used to assemble the organism, a problem
that remains, a hundred years after its recognition, beyond our
present capabilities.

It was at the end of the meeting, however, that a certain
realization surfaced, a truth that only underscored the isolation
it is possible to achieve outside the world of politics and business. Although assembly of the organism may continue to be
the outstanding biological question of the next twenty years,

if our present cultural activities continue in roughly the same manner as they are manifest today, the Earth of 2004 will have to support two to four billion additional people. This population growth itself would not be so much of a problem were it not likely to be accompanied by other global trends: These additional two to four billion people will probably be underfed, undereducated, without adequate health care, and in possession of an additional fifteen trillion dollars' worth of new weapons. One wonders what each poverty-stricken, malnourished, infectious-disease-ridden child will do with his or her respective four billion dollars' worth of new missiles, small arms, and ammunition. One wonders how many of them will be able to discuss at any length the intrinsic worth of *H. sapiens,* a single species of animal.

Who, finally, has the responsibility to deal with the factors that are even now contributing to produce the situation just described? In whose domain lies this phenomenon in which a single species is generating for itself an environment that can only partially be described by physical measurements? A select few biologists have completed their careers with thoughts on the human condition. But the security of success in the lab has almost seemed to be a prerequisite for this final metamorphosis. My question to the young scientist who will hit mid-career in 2004 is: Must one have a Nobel Prize, like Peter Medawar, before venturing into the arena of culture? Must we be a Lewis Thomas (*Lives of a Cell*) before our thoughts are worthy public fare? Of course not. Thus, to the person becoming a biologist, I answer the question opening this section with a yes: The study of culture is not only an appropriate pursuit, it comes close to being a matter of professional responsibility. However, the earlier this feeling arrives the more dangerous it will be to your science. It is very difficult to concentrate on the construction of hypotheses, the design of experiments, the mixing of chemicals, when you've accepted responsibility for the future of *H. sapiens.*

5.

In an earlier chapter I encouraged you to study what you want to study, and offered the historical reasons to support this advice. There is, however, another, more compelling argument, one that relates to the nature of human beings, for pursuing a course of study to satisfy your own curiosity. This argument falls in the category of obligations to oneself, obligations that become apparent when you struggle, as every person should, with the questions: Who am I, what am I, what do I have to offer the world, what does my life mean? There is no need to remind you of the title of this book or that biologists as a group are a notoriously agnostic bunch. So it should be no surprise that the answers I suggest reflect an overriding humanism rather than a religious view of life.

An individual human being has talents, insights, approaches to problem solving, thoughts, visions of reality, desires, ambitions, and the ability to change or maintain certain situations. The responsibility to oneself is to recognize this fact and act upon it. Consciousness provides us with the potential for self-awareness, yet it is more common than not for people, especially young people, to actually neglect their unique abilities and potential. Our species seems at times to be so highly social that self-awareness is suppressed. Ashley Montagu and Floyd Matson address this problem in a frightening book, *The Dehumanization of Man*. Their contention is that certain social conditions effectively destroy awareness of self as human, thus destroying the feeling of moral consciousness.

From the entertainment industry, Montagu and Matson cite *The Godfather* (1972), *The Texas Chainsaw Massacre* (1976), *Halloween I, II,* and *III* (1978, 1980, 1982), and many other films as examples of violence as entertainment, a phenomenon they feel greatly affects our respect for ourselves as humans able to make moral choices. They also cite Lt. William Calley's obeying an order to destroy an entire Vietnamese village, the Charles Manson gang's submission to his every command, and Gordon Liddy's waiting for orders to kill a journalist, and contend that with the technological age has come a "robopathic" personality which is "rigidly conformist, compulsively orderly and efficient . . . and unquestioningly obedient . . ." Much of our social structure rests on authority, of course, but Stanley Milgram showed through the use of experiments that under the direction of an authoritative person in a lab coat, a volunteer "teacher" would punish a "learner" for giving wrong answers by administering electrical shocks in increasing voltages up to the point of electrocution. The "teachers" were unsuspecting subjects—there was no electricity involved; and the "learners" were trained helpers who acted as if they were receiving the shocks. Milgram concluded from this work that, in obeying instructions, ordinary citizens (his "teachers") not driven by emotion would administer a four-hundred-fifty-volt shock to a person who did not respond correctly to a question.

Why are we approaching a biologist's responsibility to himself or herself under this aegis? Because employment pressures, even those in academia, can exert powerful dehumanizing influences. Rules, regulations, policies, financial constraints, team ideas, group votes, and the assortment of imagined needs of a subdiscipline identification all threaten to erode consciousness, under the very terms of the social contract: One doesn't have to decide how or even whether to act because that act is covered by commonly accepted practices. Lewis Thomas' musings in *Lives of a Cell* on the maintenance of body functions—he is relieved not to have to make decisions for his liver—hint at the biological underpinnings of social contract. But Montagu and Matson counter that the ultimate effects of responsibility denial include some of history's most stunning acts of immorality.

But back to science. Acceptance of responsibility to oneself demands actions that acknowledge the potential for unique contributions. Put more bluntly, your own ideas, as distinct from the ideas of others, must take on a certain value, and hence become worthy of use in formulating hypotheses and theories. When you personally carry out this part of your own work, then you have made *a*, if not *the*, major step toward becoming a scientist instead of a technician, a biologist instead of one who simply makes a living doing tasks that require biological knowledge.

If we accept the premise, the salient questions become, as they must for all professions, and not just biology: Am I really bringing my unique contribution to the world, or am I just solving problems others feel merit the effort? Even if the latter is the case, one should ask: Do my solutions fall within the range of those accepted by the paradigm, or am I at least considering some resolutions that, when viewed from the perspective of history, would be called breakthroughs? These questions finally address the worth and role of the individual. My contention is that every person's ideas have worth. Unless one accepts responsibility to oneself, the full extent of this worth will never be realized.

The pursuit of truly unique ideas is rarely a smooth experience, and history has shown individuality to be a stressful way of life, often extremely so, depending on the political/intellectual environment in which it is lived. But individuality and creativity are not, in the end, absolute requirements for becoming a biologist. Indeed, one of the ways in which the professions most resemble the trades is that it is quite possible to be a practicing anything, biologist, attorney, physician, engineer, for an entire life without expressing a shred of individuality. In these cases the professional is taking refuge in the profession.

6.

Suppose biologists, as scientists, refuse to accept their responsibilities. What might be the results of such a decision? Answers range from the relatively naive "nothing" to the arrogant and elitist "the ignorant shall inherit the Earth." In order to arrive at what I hope is a sensible answer, which probably lies somewhere between these two extremes, I'll return to some human characteristics mentioned earlier: the ability to deal with different forms of reality (weights, measures, feelings, emotions, ideas, concepts) and the habit of constructing environments of the mind, or deciding that we are living a certain way regardless of what the evidence suggests. With these traits, humans are able to build an image of the world they take to be the truth. If we then consider volition as an added factor in human biology, we see a system of stimulus and response in which the stimulus can be anything from physical reality to a thought, and the response can be drawn from the whole range of human behavior, including the establishment of further forms of "truth." Translated, this statement means that in developed countries humans can easily, and perhaps most often do, live the bulk of their lives at a psychological distance from the visible and tangible world inhabited by other species.

The uniquely human world of ideas is in stark contrast to the "natural" world of radiant energy and molecules, except in one regard: The physical world provides the complex organic molecules used by humans to maintain brains that generate, somehow, the minds. *Homo sapiens*, no matter how much it believes in the possibility, still cannot escape this ultimate contact with physical reality. As a result, actions born in one world —namely, that of the mind—may have long-term consequences in another world, that of the body. Examples range from toxic chemical contamination to acid rain and the nuclear arms race. Without the reality constructed by the mind, such phenomena would not exist. A biologist, however, immediately recognizes mental products for what they are: things that have the potential for greatly altering our physical environment.

Because of his or her understanding of life processes, the scientist knows, or at least can guess, the eventual effects of these environmental alterations. The biologist therefore serves as the source of collective awareness, the source of our concern about the consequences of our actions in the physical world. "Biologist" in this context, of course, refers to more than the professional. It may include the biologist streak in literally millions of citizens. The extent to which the lay public is, and remains, aware is the extent to which biological scientists continually refresh the public memory.

Perhaps I am overstating my case, but I remain convinced that if biologists do not assume responsibility for public awareness, for the consequences of actions that affect the common physical world, then no one else will. Whatever awareness that does exist will consequently disappear and the biosphere will progress steadily toward an uninhabitable state.

A biologist's responsibility is an awesome one. To bear it demands communication skills that are never taught in biological science classes, but are instead acquired through a conviction that they are necessary. Such skills are developed largely by regular and serious contact with lay people, often in the freshman classroom. Thus no matter how heady the temptation, a young biologist cannot afford to take refuge in the cloistered lab and circle of excited friends who share a devotion to a narrow range of biological subjects. It is sad, of course, to imagine a young scientist who does not have a coterie of associates with common interests. In science, as in many other fields, the jostling of ideas often seems to generate sparks of insight that cannot be predicted from a study of individual contributions. But your biologist friends are not the world. Instead, they are supported by the world. They operate in a society maintained largely by farmers, by businessmen and -women, by those who extract natural resources—in effect by those driven mostly by money and the sense of security that comes with employment. These people often see an ecological consciousness as antithetical to their way of life. Such a vision may have short-term validity. It is, however, long-term idiocy.

Every biologist knows this judgment to be true. For example, it is not difficult to calculate long-term results of certain processes, especially those that involve growth. Dr. Al Bartlett, a physicist at the University of Colorado, has made a religion of such calculations and devoted his later years to spreading the gospel. With sobering calmness he evaluates the number of U.S. citizens who should be able to understand the mathematics of growth. The number includes all those with the equivalent of high school algebra: all scientists, engineers, physicians, accountants, stockbrokers, bankers, most businessmen and -women, vast thousands of students and educated people in other occupations, including, one would think, politicians. Bartlett then lists areas in which we seem to have accepted growth as an inevitable process: wages, prices, taxes, population, fossil fuel consumption, energy production, "business," gross national product. Finally, he asks: What is a generally acceptable level of growth? If your salary increases 5 percent per year, is that sufficient? Is 5 percent per year inflation tolerable? Although a 5 percent increase in most areas seems disappointing, for the purpose of discussion let us say most people are willing to accept it. Yes, a 5 percent per year growth rate, across the board, seems fine.

A child born in 1985 will pay $24,000 for his or her first new car at age sixteen if prices increase five percent annually. If it is driven 10,000 miles per year for fifteen years, a second new car to replace it will cost $50,000. Such increases seem reasonable: Our earnings, after all, are supposed to grow with prices. Wages and prices are relative. Within our frame of reference, however, humans are absolute entities whose numbers, at least for the present, are growing. At 5 percent per year growth in population, rural Kansas towns turn into Chicagos over the next lifetime. There are many scientists like Dr. Al Bartlett who can envision a time at which, theoretically, the mass of humanity will equal the mass of the Earth.

The time is not so distant in historical terms. Today we study classical Greek myths, view museum artifacts from ancient Persian Gulf cultures, speak a language whose roots lie in extinct Rome. A significant fraction of us believe, with no proof but the written word, that Jesus walked the Earth as corporal God two thousand years ago. Picture yourself in the dusty streets of biblical Nazareth. It is not a difficult exercise: Films, books, weekly services, and a string of religious holidays keep us in touch with the Roman Empire of two millennia past, we think. But try to share the realities of your own era with these people. Explain a computer, molecular biology, evolution, television, the Apollo moonshots (remember them, the greatest of our national scientific ventures?), nuclear weapons. Tell them the Earth revolves around the sun. Describe the rings of Saturn, quasars, black holes, subatomic particles, automobiles. Tell them whales sing.

What is their reaction? Is it the same as yours would be today when asked to see through time? Have humans, collectively, in order to effect a rational decision, ever been able to envision a world beyond their immediate experience? I don't believe they have. The world suffers from desertification; we concern ourselves with a college education and a job. The world gives us war and oppression; we foresee a marriage and, perhaps, children. The world destroys tropical forests and dumps nuclear and toxic wastes into everyman's food; we look for excitement in the cities. Do we imagine a time when petroleum is gone, when the major nations take up clubs against one another, when the mass of humanity equals the mass of the Earth? Probably not. Humans will ultimately have two options: to live the life the technological world they created offers them, or to do something that in modern times has proven exceptionally difficult—namely, to make a collective rational decision, based on present knowledge, to live within the resources provided by Earth.

With deadly calmness, Dr. Bartlett continues his analysis with a list of tools from which humanity can choose to shape its future. The list has two categories:

A. Those phenomena that, on a global scale, exacerbate our problems of overpopulation; food, energy, and water shortage; psychological stress:

Motherhood
Babies
Large families
Medical care
Medical research
Health
Modern agriculture
55 MPH speed limits
Peace
Auto seat belts and air bags
Smoking bans

B. Those phenomena that, on a global scale, help solve our problems of overpopulation; food, energy, and water shortage; and psychological stress:

Contraception
Abortion
Homosexuality
War
70 MPH speed limits
Smoking
Cancer
Famine
Infectious disease
Infanticide
Euthanasia

Al Bartlett's conclusion is that our problems will, in fact, be solved before the mass of humanity comes to equal the mass of the Earth. To effect that solution, we may either choose from our list of known devices, or we may decide to let the choice be made for us. The most rational item on list B, contraception, is still, on a worldwide basis, exceedingly controversial, and abortion is an order of magnitude more so. A scientist viewing us as he or she would a natural population of some nonhuman species, for example ants, would conclude that we have already selected the fourth item on list B as our solution, and are able to start this population control mechanism almost at will. And, when we're not actually doing it, we're preparing for it.

Is this state of affairs a scientific, biological issue? Yes. As a biological scientist, you have a hierarchy of responsibilities, not the least of which is to bring your understanding to a world that, without it, will make choices that negate the values for which you stand: the values that derive from a naturalist tradition, a reverence for organisms, and a commitment to a lifetime of learning and teaching.

ANNOTATED BIBLIOGRAPHY

AND

READING LIST

Periodically students ask me for suggestions about books to read. Such questions usually occur after I've mentioned something in class; it is evidently strange to them to see a person so affected by books. So over the years I've developed a list which I simply hand to them with the request that they come back after they've waded through a couple of the titles. The following list includes some of these books as well as my comments about them. Although a few may seem strange ones to support *On Becoming a Biologist,* many of them strongly influenced my view of the profession because their authors were able to reflect on questions with which I had been wrestling. Perhaps some of these people were able to chip away the narrowness with which I happened to be seeing the world at the time.

The list also includes some titles that are either used as sources for, or referenced in, the text, but are not on my student reading list. The two categories are mixed. The reading list, in practice, is changed almost daily.

Adler, Mortimer J. *The Paideia Proposal.* New York: Macmillan, 1982. 84 pages. Adler's statement for the Paideia Group about the components and directions of education that are requisite for individual development and an enlightened free society. This book is probably the most concise argument in print for the value of reading, writing, and mathematics in human endeavors.

Baltimore, David. *Nobel Lectures in Molecular Biology.* New York: Elsevier, 1977. 534 pages. Essays by, and short biographies of, Nobel Prize winners, including Baltimore himself.

Cavalli-Sforza, L. L., and M. W. Feldman. *Cultural Transmission and Evolution: A Quantitative Approach.* Princeton, N.J.: Princeton University Press, 1981. 388 pages. A highly stimulating and thought-provoking (although quite mathematical) analysis of the movements of ideas through societies, this book has probably done more to influence my approach to teaching than anything else I have read.

Darwin, Charles. *The Autobiography of Charles Darwin.* New York: Harcourt, Brace & World: 1958. 253 pages. This is the "first complete edition" with notes, replacement of parts originally omitted, and an appendix by Darwin's granddaughter Nora Barlow. Darwin's description of his career provides us with a portrait that is simply not available from secondary sources, and includes, for example, his lament over the atrophy of his love for literature.

Durant, Will and Ariel. *The Lessons of History.* New York: Simon and Schuster, 1968. 117 pages. The Durants' chapter on the lessons of biology probably summarizes as well as any paper the accomplishments of the field.

Dyson, Freeman J. *Disturbing the Universe.* New York: Harper & Row, 1979. 283 pages. Dyson looks back on his lifetime of scholarship, and the careers of those around him, in a way that any scientist can appreciate because it deals with both the personal and political impact of discoveries.

Eiseley, Loren. *All the Strange Hours.* New York: Scribner, 1975. 273 pages. Eiseley was basically a naturalist who was evidently troubled, almost to an obsession at times, by his ability to "see through time." His autobiography reveals as well as can be shown a scholar's struggle to explain the meaning of his life and the significance of his contributions.

Eiseley, Loren. *Darwin's Century.* Garden City, N.Y.: Doubleday, 1958. 378 pages. An examination of the intellectual climate in which the theory of natural selection developed, this work includes some interesting appraisals of the influence of various eighteenth- and early-nineteenth-century scientists on Darwin.

Farb, Peter. *Man's Rise to Civilization.* New York: Dutton, 1968. 332 pages. Farb gives us a popular version of cultural evolution, eloquently written and convincing, which reflects on the manner in which our present society deals with innovation and complexity, both matters of concern to scientists.

Farb, Peter. *Word Play: What Happens When People Talk.* New York: Knopf, 1973. 350 pages. A useful lesson in language for the nonlinguist, this book makes one acutely aware of both the potentials and problems of trying to communicate, no matter what the subject matter.

Futuyma, Douglas J. *Science on Trial—The Case for Evolution.* New York: Pantheon Books, 1983. 251 pages. An exceptionally articulate analysis of the differences between evolutionary biology and creation science, this book also serves to characterize science in general and biology in particular.

Gall, John. *Systemantics.* New York: Quadrangle/New York Times, 1975. 111 pages. A tongue-in-cheek look at our institutions and organizations, Gall's essay would be a great deal funnier if it were not so depressingly true. I regularly lend my copy of *Systemantics* to administrators; it has yet to do much good.

Greenberg, Daniel S. *The Politics of Pure Science.* New York: New American Library, 1968. 303 pages. This work, probably more than any other I have read, shows that science is as human an activity as anything else we do, and that it is not immune to the forces that drive politicians and businessmen.

Hixson, Joseph. *The Patchwork Mouse.* Garden City, N.Y.: Anchor/Doubleday, 1976. 228 pages. A sad tale about how one biologist handled the pressures of research. This book makes one look twice at the way reputation functions as the currency in science.

Hofstadter, Douglas R. *Gödel, Escher, Bach: An Eternal Golden Braid.* New York: Vintage Books, 1980. 777 pages.

Janovy, John, Jr. *Yellowlegs.* New York: St. Martin's Press, 1980. 192 pages.

Koestler, Arthur. *Bricks to Babel*. New York: Random House, 1980. 697 pages. Koestler's writings are characterized by a brutally objective look at many of the events that have shaped a number of societies and cover an exceptionally broad range of subjects, including ones that are of major interest to scholars.

Koestler, Arthur. *The Case of the Midwife Toad*. New York: Random House, 1971. 187 pages. The ultimate study of academic politics surrounding a biological idea that refuses to die.

Kuhn, Thomas S. *The Structure of Scientific Revolutions*. Chicago: University of Chicago Press, 1970. 210 pages.

LeShan, Lawrence, and Henry Margenau. *Einstein's Space and Van Gogh's Sky*. New York: Macmillan, 1982. 268 pages.

Lorenz, Konrad. *On Aggression*. Harcourt, Brace & World, New York: 1966. 306 pages.

Mailer, Norman. *Of a Fire on the Moon*. Boston: Little, Brown, 1970. 472 pages. Perhaps one of the best descriptions of modern humanity's interactions with high technology, this book is important reading for students because of its reminder that we once set out on an unprecedented scientific adventure (manned space travel) that touched all our lives, were successful at it, and then for some reason abandoned it.

Mead, Margaret. *Blackberry Winter*. New York: Morrow, 1972. 305 pages. Regardless of how anthropologists and others ultimately evaluate her work, Mead's story here of how she became involved in scholarship has much to say to a young person contemplating a career that doesn't seem to offer large financial rewards.

Melville, Herman. *Moby Dick*. New York: Macmillan, 1962 edition. 621 pages. I strongly recommend a rereading of *Moby Dick* from the perspective of a biologist; it reveals attitudes toward nature that are prevalent today, and Melville's story is made all the more powerful by our present knowledge of whales.

Montagu, Ashley, and Floyd Matson. *The Dehumanization of Man*. New York: McGraw-Hill, 1983. 246 pages.

Pirsig, Robert M. *Zen and the Art of Motorcycle Maintenance*. New York: Morrow, 1974. 412 pages. A study of values and approaches to life, especially those of the university teacher, that has become a virtual cult piece. I know scholars who pass judgment on people based on whether they have read this book!

Rothschild, Miriam. *Dear Lord Rothschild*. Glenside, PA: Balaban Publishers, 1983. 398 pages.

Shine, I., and S. Wrobel. *Thomas Hunt Morgan: Pioneer of Genetics*. Lexington, KY: University of Kentucky Press, 1976. 160 pages.

Steinbeck, John. *The Log of the Sea of Cortez*. New York: Viking Press, 1941. 598 pages. This book is a rare contribution in that it gives us a view of nature, and of a biologist, by a winner of the Nobel Prize for Literature.

Sutton, George M. *Bird Student*. Austin, TX: University of Texas Press, 1980. 216 pages. Sutton's experiences with Fuertes offer a major case study of the relationship between a world-renowned naturalist and a person who chose to become a biologist at an early age. The two Sutton books complement one another; the letters include captivating advice about how to practice one's craft, while the autobiography tells what Sutton did with that advice.

Sutton, George M. *To a Young Bird Artist*. Norman, OK: University of Oklahoma Press, 1979. 147 pages.

Thomas, Lewis. *Lives of a Cell*. New York: Viking Press, 1974. 153 pages. My copy is virtually worn out from being read and reread, having passages marked, and being carried on airplanes.

Tuchman, Barbara. *Stillwell and the American Experience in China*. New York: Macmillan, 1970. 621 pages. Tuchman presents an exceptionally clear description of the encounter between two cultures, revealing in the process that the distance between cultures on Earth can be as great as that often described in science fiction. Thus the book serves as a model for the encounter between a scientist and the general population.

Turnbull, Colin M. *The Mountain People*. New York: Simon and Schuster, 1972. 309 pages. A discussion of what happens to people when their lives are reduced to subsistence level and competition over food—that is, a condition devoid of values and arts. I recommend this book to students who are too interested in a college experience that will give them a "good job."

Watson, James D. *The Double Helix.* New York: Atheneum, 1968. 226 pages. Watson describes his own work surrounding a discovery that not only won the Nobel Prize, but also changed forever the nature of biological research.

Wilson, Edward Osborne. *On Human Nature.* Cambridge, MA: Harvard University Press, 1978. 260 pages. It is important to read not only some of Wilson's basic thoughts on biology, but also his statements on humanity that are derived from his behavioral research on social animals.

Wilson, Edward Osborne. *The Insect Societies.* Cambridge, MA: Harvard University Press, Belknap Press, 1971. 548 pages.

Zinsser, Hans. *Rats, Lice and History.* Boston: Atlantic/Little, Brown, 1934. 301 pages. Zinsser relates the classic tale of interactions between nature and humanity, especially as they influence military actions and therefore subsequent political events.

Zukav, Gary. *The Dancing Wu Li Masters.* New York: Morrow, 1979. 352 pages. This book is important because it not only tells how basic research generates discoveries that give us a certain world view, but it also suggests that our perceptions of what the world is really like may need considerable reevaluation.